Audel™

Carpenters and Builders Tools, Steel Square, and Joinery

Audel™

Carpenters and Builders Tools, Steel Square, and Joinery

All New 7th Edition

Mark Richard Miller
Rex Miller

Wiley Publishing, Inc.

Vice President and Executive Group Publisher: Richard Swadley
Vice President and Publisher: Joseph B. Wikert
Executive Editor: Carol A. Long
Editorial Manager: Kathryn A. Malm
Development Editor: Kevin Shafer
Production Editor: Angela Smith
Text Design & Composition: TechBooks

Copyright © 2005 by Wiley Publishing, Inc., Indianapolis, Indiana. All rights reserved.

Published by Wiley Publishing, Inc., Indianapolis, Indiana
Published simultaneously in Canada

For general information on our other products and services please contact our Customer Care Department within the United States at (800) 762-2974, outside the United States at (317) 572-3993 or fax (317) 572-4002.

Wiley also publishes its books in a variety of electronic formats. Some content that appears in print may not be available in electronic books.

Library of Congress Control Number:

ISBN: 0-7645-7115-X

Printed in the United States of America

10   9   8   7   6   5   4   3   2   1

Contents

Foreword xiii

Acknowledgments xv

About the Authors xvii

Introduction xix

Chapter 1	Technical Drawing	1
	Mechanical Drawing	2
	Mechanical Drawing Instruments	2
	Mechanical Drawing	13
	Electronic Drawing	25
	Computers	26
	Why Use CAD?	29
	The CAD System	30
	CAD Concepts	31
	A Simple Drawing	35
	Plotting	41
	Summary	42
	Review Questions	43
Chapter 2	Safety	45
	Clothing	46
	Protective Equipment	47
	Hand Tools	47
	Power Tools	48
	Circular Saw Safety	48
	Table Saw Safety	49
	Radial Arm Saw Safety	49
	Portable Electrical Drill Safety	50
	Other Power Tools	51
	Good Housekeeping	51
	Excavations	51
	Scaffolds and Ladders	51
	Decks and Floors	53
	Falling Objects	53

	Lifting and Carrying	53
	First-Aid	53
	Summary	54
	Review Questions	54
Chapter 3	**Guiding and Testing**	55
	Straightedge	55
	Square	56
	Miter and Combined Try-and-Miter Squares	59
	Framing or Steel Square	62
	Combination Square	66
	Sliding T Bevel	67
	Center Square	70
	Level	70
	Plumb Bob	70
	Summary	72
	Review Questions	72
Chapter 4	**Layout Tools**	75
	Chalk Box and Line	75
	Carpenter's Pencil	75
	Ordinary Pencil	76
	Marking or Scratch Awl	76
	Scriber	76
	Compass and Dividers	79
	Summary	80
	Review Questions	80
Chapter 5	**Rules, Scales, and Gages**	81
	Rules	81
	Folding Wood Rule	81
	Other Rules	82
	Using Rules	83
	Lumber Scale	84
	Spring Steel Board Rule	85
	Marking Gages	86
	Single-Bar Gage	86
	Double-Bar Gage	86

	Slide Gage	87
	Butt Gage	88
	Summary	89
	Review Questions	90
Chapter 6	**Clamps, Vises, and Workbenches**	**91**
	Horses or Trestles	91
	Clamps	91
	C-Clamp	91
	Deep-Throat Type	92
	Edge Clamp	94
	Pipe Clamp	94
	Steel-Bar Clamp	94
	Spring Clamp	94
	Hand Screw	94
	Band Clamp	94
	Vises	96
	Woodworker's Vise	96
	Bench Vise	97
	Clamp-On and Sawhorse Vises	98
	Workbenches	98
	Bench Types	99
	Features	100
	Summary	105
	Review Questions	105
Chapter 7	**Saws and Sawing**	**107**
	Saw Characteristics	107
	Types	107
	Saw Teeth	109
	Set	109
	Action of the Crosscut Saw	109
	Action of the Ripsaw	111
	Angles of Saw Teeth	111
	Coping Saw	111
	Summary	114
	Review Questions	114

Chapter 8	**Chisels, Hatchets, and Axes**	**115**
	Chisels	115
	Paring Chisel	116
	Firmer Chisel	117
	Framing or Mortise Chisel	117
	Slick	117
	Note	117
	Gouge	117
	Tang and Socket Chisels	117
	Butt, Pocket, and Mill Chisels	117
	How to Select Chisels	118
	How to Care For and Use Chisels	120
	How to Sharpen Chisels	121
	Drawknife	124
	Hatchets and Axes	125
	Broad Hatchet or Hand Axe	125
	Axe	125
	Adze	128
	Summary	129
	Review Questions	130
Chapter 9	**Planes, Scrapers, Files, and Sandpaper**	**131**
	Spoke-Shave	131
	Planes	132
	Jack Plane	134
	Fore Plane	134
	Jointer Plane	134
	Smoothing Plane	134
	Block Plane	135
	Rabbet Plane	136
	Surform	136
	Grooving Plane	137
	Router	137
	Plane Irons or Cutters	138
	Bevel of the Cutting Edge	141
	Double Irons	143

	Plane Mouth	143
	How to Use a Plane	143
	Scrapers	147
	Files and Rasps	151
	Sandpaper	152
	Summary	154
	Review Questions	155
Chapter 10	**Awls, Augers, Bits, and Braces**	**157**
	Awls	157
	Augers	157
	Twist Bits	162
	Countersinks	162
	Hand Drills and Braces	163
	Using Boring Tools	165
	Summary	165
	Review Questions	166
Chapter 11	**Hammers, Screwdrivers, Wrenches, and Staplers**	**167**
	Hammers	167
	Screwdrivers	172
	Wrenches	174
	Other Fastening Tools	176
	Summary	177
	Review Questions	178
Chapter 12	**Portable Power Tools**	**179**
	Cordless Power Tools	179
	Advantages	180
	Disadvantages	180
	Standard Electric Tools	181
	Electric Drills	182
	Portable Saws	182
	Electric Planes	183
	Saber Saws	184

	Sanders	186
	Routers	186
	Summary	189
	Review Questions	190
Chapter 13	**Tool Sharpening**	**191**
	Grinding	191
	Bench Grinders	193
	Operation	194
	Safety Precautions	194
	Oilstones	194
	Honing	194
	Using Oilstones	195
	Types of Oilstones	198
	Natural Oil Stones	198
	Artificial Oil Stones	199
	Summary	200
	Review Questions	200
Chapter 14	**Saw Blade Sharpening**	**201**
	Jointing	202
	Shaping	202
	Setting	204
	Filing	205
	Dressing	208
	Summary	208
	Review Questions	208
Chapter 15	**Cabinetmaking**	**209**
	Tools	209
	Joints	209
	Glued Joints	211
	Beveled Joints	211
	Hidden Slot Screwed Joints	212
	Coopered Joints	213
	Plain or Butt Joint	213
	Straight Plain Edge Joint	214
	Dowel Joints	215

	Square Corner Joint	218
	Mitered Joints	219
	Splined Joint	222
	Splice Joints	224
	Lap Joints	226
	Rabbet Joint	226
	Dado Joint	227
	Scarf Joints	227
	Mortise-and-Tenon Joints	231
	Dovetail Joints	243
	Dovetail Angles	248
	Halved Lap and Bridle Joints	252
	Construction	254
	Framed Construction	254
	Carcass Construction	255
	Plate Jointery	256
	The System	257
	Glue Application	260
	Plates	260
	Assembly	260
	Advantages and Disadvantages	260
	Summary	261
	Review Questions	262
Chapter 16	**Mitering**	**265**
	Miter Tools	265
	Moldings	267
	Mitering Flush Moldings	268
	Mitering Spring Moldings	268
	Mitering Panel and Raised Moldings	271
	Cutting Long Miters	273
	Coping	274
	Summary	275
	Review Questions	275
Chapter 17	**Using the Steel Square**	**277**
	Application of the Square	278

Scale Problems	278
Square-and-Bevel Problems	284
Note	285
Table Problems	289
Main or Common Rafters	289
Hip Rafters	291
Valley Rafters	293
Jack Rafters	295
Cripple Rafters	295
Finding Rafter Lengths Without the Aid of Tables	295
Rafter Tables	300
Reading the Total Length of the Rafter	300
Reading Length of Rafter per Foot of Run	307
Table of Octagon Rafters	309
Table of Angle Cuts for Polygons	311
Table of Brace Measure	312
Octagon Table or Eight-Square Scale	314
Essex Board Measure Table	314
Summary	315
Review Questions	316
Index	**317**

Foreword

The *Audel Carpenters and Builders Tools, Steel Square, and Joinery: All New Seventh Edition* is one of four volumes that cover the fundamental tools, methods, and materials used in carpentry, woodworking, and cabinetmaking.

This volume was written for anyone who wants (or needs) to know about the proper use of tools used by carpenters in construction, woodworking, or cabinetmaking. Whether remodeling an existing home or building a new one, the rewards of a job well done are many-fold.

This book has been prepared as a practical guide to the selection, maintenance, installation, operation, and repair of wooden structures. Carpenters and woodworkers (as well as cabinetmakers) should find this book (with its clear descriptions, illustrations, and simplified explanations) a ready source of information for the many problems that they might encounter while building, maintaining, or repairing houses and furniture. Both technical and nontechnical persons who want to gain knowledge of woodworking and house building will benefit from the theoretical and practical coverage in this book.

This is the first of a series of four books in the Carpenters and Builders Library designed to provide you with a solid reference set of materials that can be useful both at home and in the field. Other books in the series include the following:

- *Audel Carpenters and Builders Math Plans and Specifications: All New Seventh Edition*
- *Audel Carpenters and Builders Layout, Foundation, and Framing: All New Seventh Edition*
- *Audel Carpenters and Builders Millwork, Power Tools, and Painting: All New Seventh Edition*

No book can be complete without the aid of many people. The acknowledgments that follow mention some of those who contributed to making this up to date with current design and technology available to the carpenter. We trust you will enjoy using the book as much as we did writing it.

Mark R. Miller
Rex Miller

Acknowledgments

No book can be written without the aid of many people. It takes a great number of individuals to put together the information available about any particular technical field into a book. Many firms have contributed information, illustrations, and analysis of the book.

The authors would like to thank every person involved for his or her contributions. Following are some of the firms that supplied technical information and illustrations.

American Plywood Association
American Standard Inc.
Arrow Fastener Co.
Black and Decker
Brink and Cotton
Durable Goods
Generic Software
Hand Tool Institute
McGuire-Nicholas Co.
Milwaukee Electric Tool
Porter-Cable
Sears, Roebuck, and Co.
Stanley
Steiner Lamello AG, Switzerland
Vaughn and Bushnell

About the Authors

Mark Richard Miller finished his BS degree in New York and moved on to Ball State University, where he obtained the Masters degree. He went to work in San Antonio. He taught high school and went to graduate school in College Station, Texas, finishing the Doctorate. He took a position at Texas A&M University in Kingsville, Texas, where he now teaches in the Industrial Technology Department as a Professor and Department Chairman. He has coauthored 11 books and contributed many articles to technical magazines. His hobbies include refinishing a 1970 Plymouth Super Bird and a 1971 Road-Runner.

Rex Miller was a Professor of Industrial Technology at The State University of New York, College at Buffalo for more than 35 years. He has taught at the technical school, high school, and college level for more than 40 years. He is the author or coauthor of more than 100 textbooks ranging from electronics through carpentry and sheet metal work. He has contributed more than 50 magazine articles over the years to technical publications. He is also the author of seven civil war regimental histories.

Introduction

A carpenter should have all the tools necessary to do the work required.

Buy the Best

Buy only the tools you need and buy the best.

Acquiring tools only as you need them will give you time to learn how to buy tools. If you make a few mistakes early in your career, you will make better choices later on.

Your knowledge of the work you do determines the tools you need. If you don't know how to use a tool, there is not much practical sense in owning it.

This book on tools provides you with basic information on the most common carpentry tools. For additional information, ask the lead carpenter on your project to show you his or her tools and to help you write up a list of what you will need. This will ensure that your toolkit fulfills the requirements of the work at hand and fits in with the work style of your fellow workers.

Add to your kit of tools as you learn new methods and types of work (see Figure I-1).

Which tools are best? In selecting tools, it is important to buy only the best, regardless of cost. Select from standard makes. Examine each tool carefully to be sure there are no visible defects, but you can only tell so much from outward appearances. A poorly tempered steel tool may be discovered only through use, so buy only from a manufacturer or dealer who will replace a defective tool. Carpenters who use tools on a daily basis to earn a living know a lot about those tools. Ask for advice.

Tool Ownership and Use

Hand tools are really an extension of your body. They give you the power to do things you can't do with your bare hands. This is why carpenters have very personal feelings about their tools.

More subtle differences become very important.

Carpenters get used to working with their own tools and prefer not to have any one else fooling with them. The practical side to this is that a tool should be right where you expect it to be, right when you need it (see Figure I-2). If someone has borrowed a tool, it won't be there. Even if it was returned, but the cutting edge was a little dull or it was put in your tool belt backwards, it is an annoyance. A tool returned in poor condition becomes a major distraction and gets in the way of doing your best work. Some carpenters keep

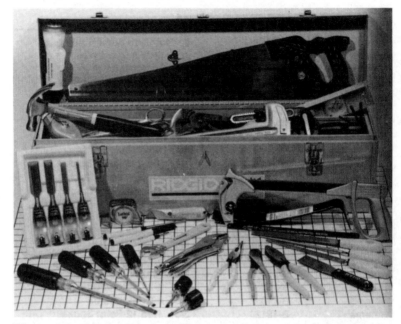

Figure I-I Nucleus tool set. This kit provides a core set of tools for the carpenter and builder that can be expanded to meet particular needs. *(Courtesy Durable Goods)*

a separate set of tools just for loaning out. This keeps their own tools at hand and helps satisfy social obligations that are difficult to avoid.

Tool Use

Use tools only for their intended purposes. Using a sharp chisel to open paint cans means it won't perform well the next time you pick it up to cut wood.

Care for tools when you're not using them. When you're finished with a tool, clean it off, sharpen it, and put it away. Then it will be ready for use the next time you need it. Storing tools in a box makes it easy to move them around and you will always have the tools you need without having to hunt.

Consider this philosophy of tool ownership and use it for an example as you develop your own.

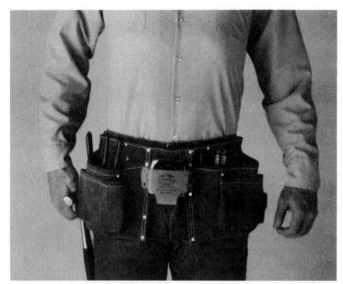

Figure I-2 A tool belt keeps tools immediately at hand as you move around the construction site. *(Courtesy McGuire-Nicholas Co.)*

Chapter 1

Technical Drawing

This chapter focuses on drawing tools, equipment, and techniques useful to the carpenter or builder. *Drawing* is the act of representing real-world objects with lines on paper. Carpenters and builders need to develop and use this skill for three main reasons.

The first is to give definite form to your own ideas. Out on the work site, you might need to cut a complex joint in framing timbers. You have an idea of how the work should go, but you are just not sure how to start. Therefore, you sketch out the joint on the back of a board to see how the parts will fit together. Laying out your idea helps you visualize the joint.

On a complex project, such as building a full set of custom kitchen cabinets, build the whole project in your mind before doing any work. This helps you locate problems and find solutions before any wood or time is wasted. However, there are dozens of details, so keep notes on paper, in the form of drawings, to help keep track of it all.

Think of these as working drawings. You make them for your own use to help get the work done.

The second reason is to document a project. You might want to build a similar project in the future. You wouldn't expect to remember all the details of building a complex project, such as a whole house. Setting the details down on paper is an aid to your memory.

The third reason for making drawings is to communicate with others. Words are inadequate to describe the complex spatial relationships that are necessary for effective communications in the building trades. In fact, depending on words alone usually leads to misunderstandings. The more said with drawings, the more complete the understanding.

When you are having difficulty describing the details of a project to an owner, a quick sketch can make everything clear.

An architect provides comprehensive drawings of a project. The drawings might be comprehensive, but they are not necessarily complete. A frequent requirement is for the carpenter or builder to submit shop drawings to the architect. These are formal drawings with more details than the architect's showing exactly how you intend to build some particular aspect of the project.

Sketches and drawings don't have to be perfectly made on the first try. The word drafting comes from the common practice of going through several drafts, refining the drawing at each stage. Going from rough sketches to freehand drawings to formal scaled drawings gives you the time and insight needed to develop meaningful drawings. Use tracing paper or transparent overlays to make changes and improvements.

Mechanical Drawing

By definition, the term *mechanical drawing* means drawing done with the aid of drawing instruments, as opposed to work done freehand. The first subject to consider is the drawing instruments themselves, and then we look at how to use them.

Mechanical Drawing Instruments

A drafter should have good instruments. In fact, the best are none too good and are easily rendered unfit for use unless they are properly handled. Unfortunately, the advice given by some instructors is to buy an inexpensive set of instruments for use until you find out if you are gifted in the art of drafting. However, if an experienced drafter cannot do good work with poor instruments, how can beginners be expected to accomplish anything or determine if they have any talent for drawing?

Many patterns and types of drawing instruments are available, and the beginning drafter should purchase only the best and most suitable equipment for his or her particular needs. The advice of an experienced drafter will be of great assistance to the student who is about to buy these instruments, which are a lifetime investment. With reasonable care, they should last indefinitely.

The following list comprises everything needed for general drafting work:

1 drawing board
1 set of instruments
1 T-square
2 triangles (30°-60° and 45°)
1 set of drawing scales
3 pencils (F, 3H, and 6H)
1 pencil eraser
1 sponge eraser
1 irregular curve
1 protractor

In some cases involving enlarging or reducing the size of drawings, proportional dividers are quite useful.

Drawing Board

A *drawing board* is usually made of a softwood or particleboard and should have a true edge (see Figure 1-1). Drawing boards are available in many different sizes, but in most cases a 20-inch × 24-inch board is large enough. To keep the drawing board surface free from dents and scratches, a vinyl drawing board top can be used. This covering offers excellent resiliency and provides a smooth drawing surface at all times.

Figure 1-1   A steel edge drawing board in wide use today.

In some cases, a drafting table (see Figure 1-2) is preferred over a drawing board. The table can be adjusted to any height and may be tilted and locked at any desired angle. The top can be adjusted from horizontal to full easel position.

The most convenient method for fastening the paper to the drawing board is to use masking tape. This comes in convenient rolls, and small pieces may readily be torn off as required. Some brands of tape may be used several times, some only once.

Parallel Ruling Straightedge

The *parallel ruling straightedge* (see Figure 1-3) is an excellent alternative to the T-square. The device is mounted over the entire drawing board. Both ends are fastened to light metallic cords that run over a system of small pulleys in such a manner that in no matter what position the straightedge is located, it is always parallel with the lower edge of the board.

Figure 1-2 A typical drawing table. The table may be tilted and locked at any desired angle.

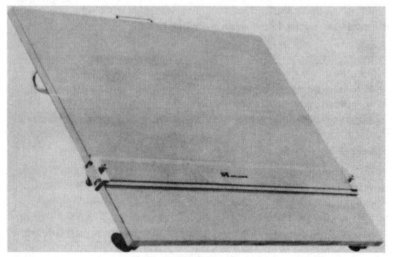

Figure 1-3 A parallel straightedge attached to a drawing board.

Drafting Machines

A *drafting machine* reduces drawing time by eliminating the manipulation of the T-square, triangles, scale, and protractor (see Figure 1-4). Parallel motion is achieved by steel bands revolving around machined drums. The machine is controlled entirely by the left hand, leaving the right hand free for drawing. Pressure can be applied to a control ring to change the setting on the protractor. The scale can automatically be snapped into any unit multiple of 15 degrees.

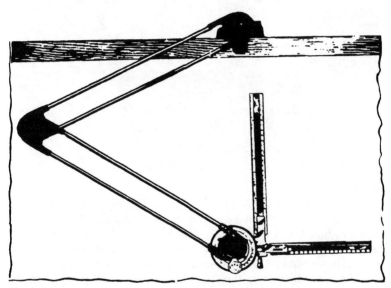

Figure 1-4 A typical drafting machine. This device can be used to take the place of a T-square, the triangles, and the protractor, since all operations that can be performed with those tools can be done with the drafting machine.

Set of Instruments

Drawing instruments are usually available in sets (that is, several instruments in a special case). For beginners and for general use, a good quality basic set containing the following instruments is all that is necessary (see Figure 1-5):

    1 compass
    1 hair-spring divider
    3 spring bows (pencil and points)
    1 extension bar

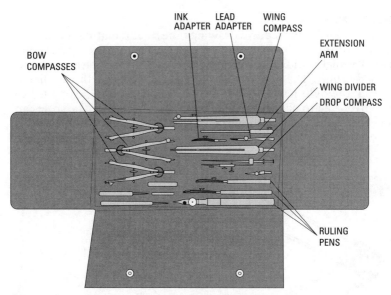

Figure 1-5 A basic set of drawing instruments.

Compass
This instrument is designed and used for describing arcs or circles. It consists of two legs pivoted together so that they may be set to any desired radius. One leg carries an adjustable needle point (or center) and the other has a joint in which the pencil arm may be secured. Each leg has a pivoted joint to permit adjustment of the ends so that the end arms (which carry the center needle point and pen or pencil) may be adjusted perpendicular to the paper for various radii (see Figure 1-6).

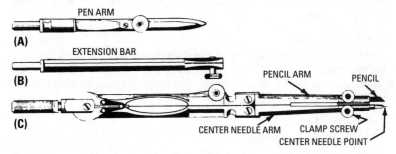

Figure 1-6 The compass with a pencil arm may also be used with a pen arm or the extension bar to inrease its versatility.

The important requirement of good compasses is that the legs have the capability of movement to any radius without any spring back. Cheap instruments may be springy, thus making it difficult to set them with precision.

Hair-Spring Dividers

Compasses and dividers are quite similar. However, each has its own special use. Dividers are instruments that have two legs pivoted at one end and sharp needle points at the other, as shown in Figure 1-7. They are used for spacing off distances. For precision, they are fitted with a hair-spring device that consists of an adjustable screw controlled by a steel spring in one leg. In operation, the legs are set to the approximate desired position and brought to the exact position by turning the adjusting screw.

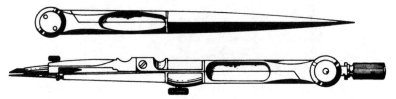

Figure 1-7   The plain and the hair-spring dividers.

Spring Bows

These small compasses and dividers are made with two spring legs whose distance apart is regulated by a small through-bolt and thumbscrew (see Figure 1-8). Spring bows are used for describing circles of small diameter and for minute spacing.

Extension Bar

To extend the range of compasses, a lengthening or extension bar (as shown in Figure 1-6B) is generally provided. This device greatly increases the diameters of circles that may be described by the compass.

T-Square

This instrument is used for drawing lines parallel to the lower edge of the board. It consists of two parts: the head and the blade. These two parts are fastened at an angle of 90° to each other, as shown in Figure 1-9. This is the fixed-head type of square. The square may also be fitted with a movable head, as shown in Figure 1-10, thereby permitting a line to be drawn at an inclined angle to the edge of the board. Figure 1-11 shows the proper positioning of the T-square.

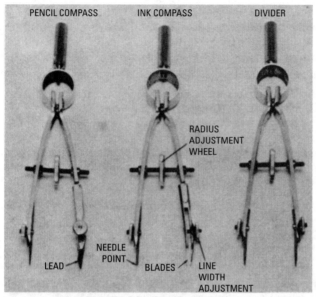

Figure 1-8 The pencil compass, the ink compass, and the dividers.

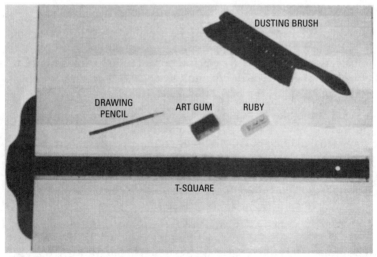

Figure 1-9 Drawing board, T-square, pencil, erasers, and dusting brush.

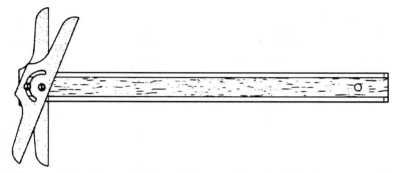

Figure 1-10 A movable-head T-square has a fixed head on one side and a movable head on the other. The movable head is pivoted so that it may be shifted to any angle with the blade.

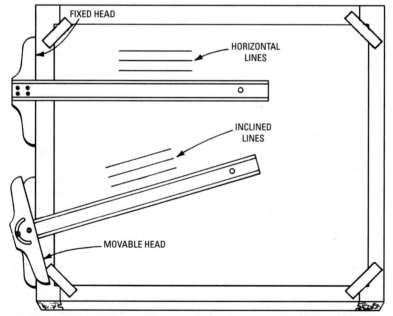

Figure 1-11 The fixed and movable-head T-squares in position on the drawing board. Horizontal lines are drawn with the fixed-head square, and inclined lines are drawn with the movable-head square.

Triangles

For drawing other than horizontal lines, triangles are generally used.

Two triangles will ordinarily be required: the 45° triangle and 30° triangle (see Figure 1-12). Figure 1-12A has two equal sides at right angles with the third side making an angle of 45° with the two legs. Figure 1-12B also has two sides at right angles, but the third side makes a 30° angle with one leg and a 60° angle with the other leg. Figure 1-13 shows the position of the triangles on the T-square.

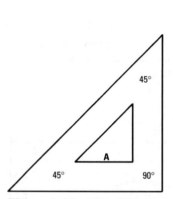 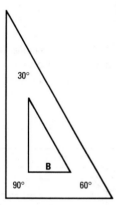

Figure 1-12 The 45° and the 30° triangles.

Rules and Scales

Scales are used to measure distances on the drawing. Good scales are made of wood faced with plastic, or they are made entirely of solid plastic. Cheaper scales are made completely of wood, and they will not long withstand hard use and constant cleaning.

When drawings are made to the same size as the object being drawn (that is, to full size), a common 1-foot rule is employed. However, when drawings are to be made larger or smaller than actual size, special scales are used. For architectural drawing, the various scales are divided into feet and inches with appropriate subdivisions (see Figure 1-14).

The triangular scale contains six different scales, as shown in Figure 1-15. Other scales are available. The scales are usually designated by the length of the foot division, as for example, the 1½ or ¾-inch scale. On each scale, the first foot is divided into inches and, where the scale is large enough, divided into fractions of an inch.

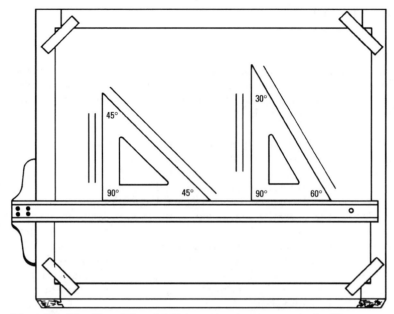

Figure 1-13 The 45° and 30° triangles in position on the T-square.

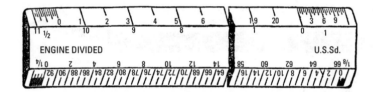

Figure 1-14 Scales used by the drafter.

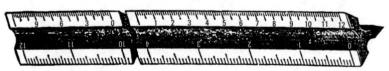

3 inches = 1 foot
1½ inches = 1 foot
1 inch = 1 foot
¾ inch = 1 foot
⅜ inch = 1 foot
¼ inch = 1 foot

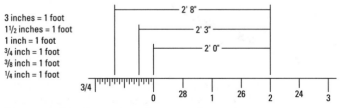

Figure 1-15 Six scales of a triangular scale.

Pencils

Drawings are generally made in *pencil*. Drawing pencils are made in various degrees of hardness, from 6B (the softest) to 9H (the hardest). The choice of hardness or softness depends on the type of drawing to be done and on the degree of precision desired. For ordinary work (such as when laying out house frames on a large scale), a 2H or 3H pencil is generally used. However, when drawing a roof-stress diagram, for example, for a scale of 1 inch = 2000 pounds, a pencil no softer than H should be used because a sharp point could not be maintained with a soft pencil, and precision could not be expected with a pencil having a blunt point.

Pencils are generally sharpened, with $3/8$ to $1/2$ inch of lead exposed, on a piece of sandpaper (see Figure 1-16), but some drafters prefer to sharpen them so the lead is wedge-shaped (see Figure 1-17). Many drafters prefer mechanical pencils, since leads of any desired hardness may be inserted instantly (see Figure 1-18). Sharpening of the leads is done on the sandpaper pad, and the leads may be used until they are quite short.

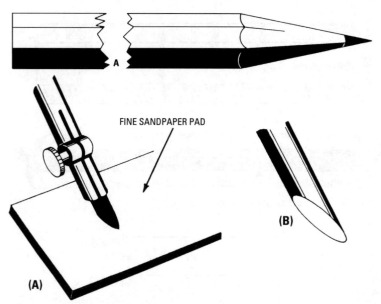

Figure 1-16 Some drafters prefer to sharpen their drawing pencils to a conical point, as in A. The compass lead is normally sharpened as shown in B.

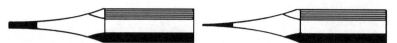

Figure 1-17 The end and side views of a drawing pencil that has been sharpened with a wedge point.

Figure 1-18 A typical mechanical pencil.

Once the lead is sharp, the loose graphite should be wiped from the lead. The secret to good line quality is to keep a sharp point on the lead.

Cleaning Powder
Keep the drawing clean while working. Use *cleaning powder* sprinkled lightly over the paper (see Figure 1-19). It readily absorbs graphite, dust, and dirt and is available in bag form or small containers.

French Curves
For describing curves other than circles, special templates called *French curves* are used to guide the pen. These may be obtained in great variety. A set of two or three will be useful. Figure 1-20 shows several forms of these curves.

Protractor
This instrument is employed for laying off or measuring angles (see Figure 1-21). Its outer edge, as shown in the illustration, is a semicircle with the center at 0. For convenience, it is divided into 180 equal parts (or degrees) from *M* to *S* and in the reverse direction from *L* to *F*. *Protractors* are often made of transparent plastic to allow the drawing under them to be seen. Figure 1-22 shows a steel precision protractor.

Mechanical Drawing
The drawing paper is stretched flat and smooth and is usually secured by means of small tabs of pressure-sensitive tape, as many as may be necessary. Use the T-square to square the paper with the board. Most drawing is done directly on tracing paper of good quality, which is normally purchased in rolls. Use pencils that are

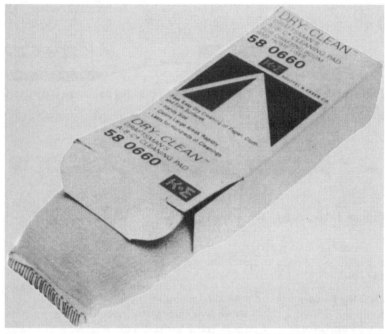

Figure 1-19 Cleaning powder used to keep drawing paper clean.

Figure 1-20 Some typical French curves.

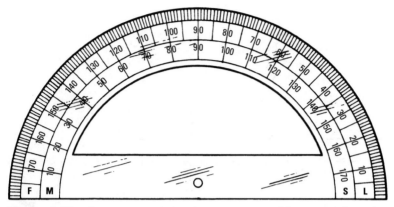

Figure 1-21 A protractor is used to measure and lay out angles.

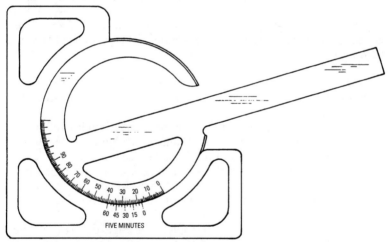

Figure 1-22 A steel precision protractor. The blade is $8\frac{1}{2}$ inches long, and graduations are read to degrees with a vernier reading to 5 minutes.

soft enough to make a good, legible black line. Prints may be made directly from the pencil drawing.

Straight Lines
To draw a straight line, use the T-square or triangle, or both, depending on the direction of the line. Horizontal lines are drawn with

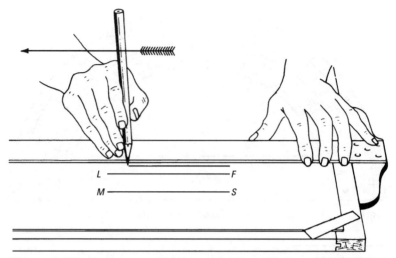

Figure 1-23 Drawing parallel horizontal lines. The lines LF and MS are drawn by moving the pencil in the direction of the arrow, guided by the edge of the T-square.

the aid of the T-square, as shown in Figure 1-23. Sometimes vertical lines are drawn by applying the head of the square to the lower, or horizontal, edge of the board.

The usual method of drawing vertical lines is with the aid of both the T-square and one of the triangles (see Figure 1-24). In this illustration, one of the legs of the triangle is used to guide the pencil. By using the hypotenuse of the 45° triangle, oblique parallel lines may be drawn (Figure 1-25), and by using the hypotenuse of the 30° triangle, oblique lines may be drawn at 30° or 60° (see Figure 1-26 and Figure 1-27).

By a combination of both triangles (see Figure 1-28), various other angles (such as 15°, 75°, 135°) may be obtained. Sometimes you may want to draw a line parallel to another line that is not inclined at any of the angles obtained with the triangles. This is done by placing the edge of one triangle parallel with the given line and sliding it along the other triangle (see Figure 1-29).

When drawing a line, it is important that the pencil be held correctly (see Figure 1-30). It should not be inclined laterally, but in drawing a line, it should be held with its axis in a plane perpendicular to the plane of the paper, slightly inclined in the direction in which it is being moved. If held as in Figure 1-30A, the inclination is likely to vary, thereby resulting in a wavy line. If held as in

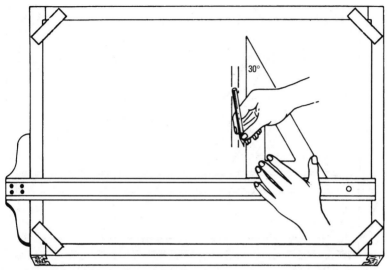

Figure 1-24   Drawing parallel vertical lines with the T-square and a right triangle. The triangle is held in contact with the T-square and is shifted to any position where a vertical line is desired.

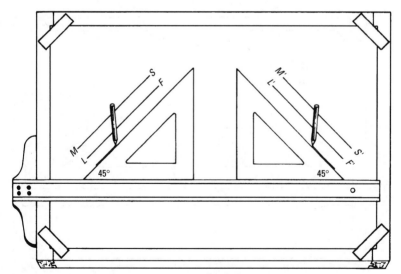

Figure 1-25   Drawing parallel 45° oblique lines with a T-square and a 45° right triangle.

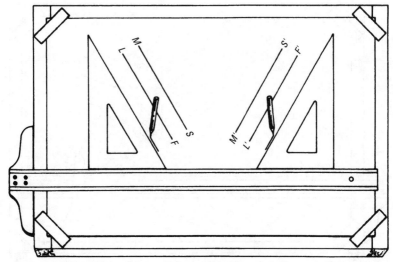

Figure 1-26   Drawing parallel 60° oblique lines with a T-square and a 30° right triangle.

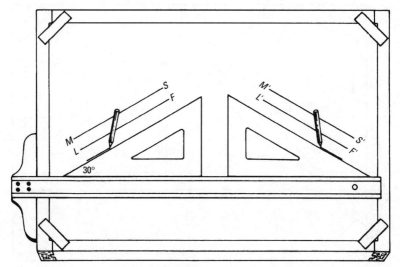

Figure 1-27   Drawing parallel 30° oblique lines with a T-square and a 30° right triangle.

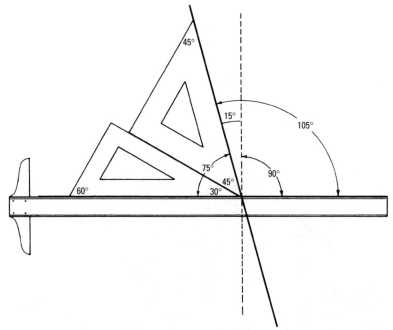

Figure 1-28   The method of drawing angles other than 30°, 45°, or 60°.

Figure 1-30B, a reference point *R* through which the line is to be drawn may not be visible or only partially visible. When held in a perpendicular plane, as in Figure 1-30C, the line comes close to the lower edge, the reference point *R* can be plainly seen, and there is only a slight chance of drawing a wavy line. When drawing lines with a pencil that has been sharpened to a conical point, the pencil should be given a slight rotation while the line is being drawn. This procedure will tend to keep the point sharp. The line will also stay the same thickness.

Arcs and Circles
These are drawn with the compass (see Figure 1-31). Both points should be approximately perpendicular to the paper, slightly inclined in the direction of movement. The starting position should be such that the entire movement can be made in one continuous sweep by grasping the handle at the pivot end by the thumb and forefinger, thus obtaining a twisting motion by moving the thumb forward without stopping to shift the hold on the compass. Never hold the

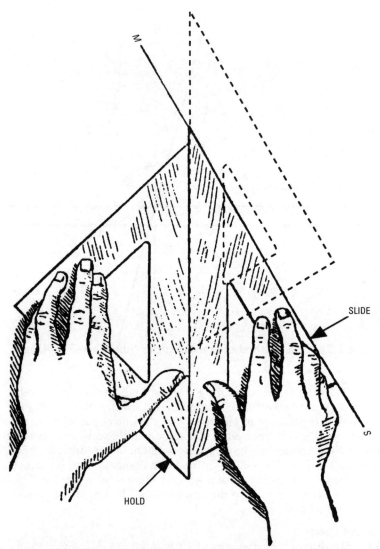

Figure I-29 To draw a line parallel to MS, slide the triangle to position R. Hold both triangles with the left hand and draw the new line.

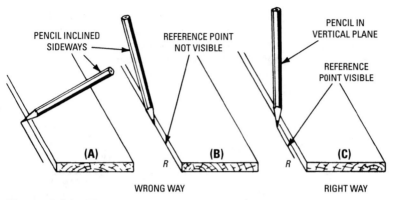

PENCIL INCLINED SIDEWAYS

REFERENCE POINT NOT VISIBLE

PENCIL IN VERTICAL PLANE

REFERENCE POINT VISIBLE

(A)

(B)

(C)

R

R

WRONG WAY

RIGHT WAY

Figure 1-30 The wrong ways and the right way of using the pencil to draw lines.

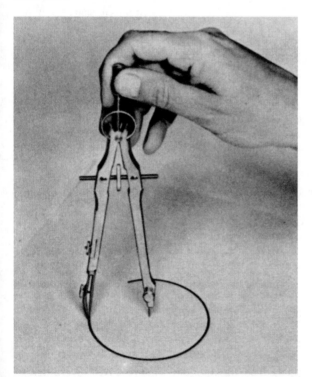

Figure 1-31 The correct use of the compass. The compass is held by the handle only; the points are nearly perpendicular to the paper.

compass by the legs, even when the lengthening bar is used. To do so will tend to move the legs to a different radius (see Figure 1-32). Always have the center point in an upright position, otherwise the indentation in the paper will be enlarged and untrue.

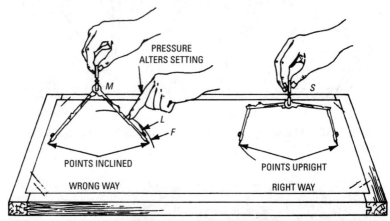

Figure 1-32 The wrong way and the right way to use the compass when describing large circles. The points, especially the needle point, should be as nearly perpendicular to the paper as possible. Use only one hand to guide the compass, since the pressure of two hands may alter the setting, and part of the circle, as at L, will vary from the true path, F.

For extremely small circles, a smaller compass (called the *bow compass*) is used. It is more convenient and, having a screw adjustment, can be set with greater precision than the large compass. Particular attention is called to the result obtained by inclining the center point of a compass in describing circles, as shown at *S* in Figure 1-33. Since these centers must be used again if the drawing is inked, accurate work cannot be done if the center indentations are spoiled (as at *S*) by the wrong or careless use of the compass.

Spacing
Hair-spring dividers are used to divide a given distance accurately into several equal parts. If an exact length is to be laid off with the dividers, a large multiple of that length should first be laid off with the scale on a right-angle line and then exactly subdivided into the desired exact length by the dividers. This involves several trials. Set the dividers as close as possible to the desired length. Then test by spacing with the dividers along the line, as shown in Figure 1-34.

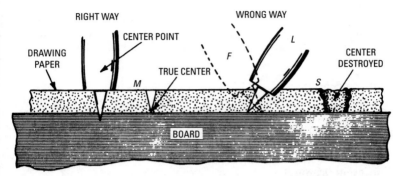

Figure 1-33 The right way and the wrong way to position the center point of a compass. The center point of a compass should be provided with a shoulder, as shown, instead of being conical; this limits the depth of the indentation.

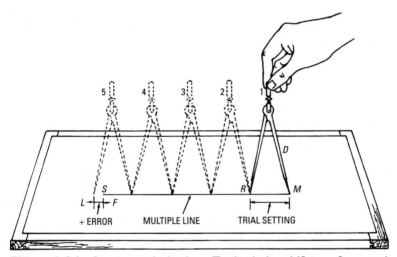

Figure 1-34 Spacing with dividers. To divide line MS into five equal parts, set the dividers by eye to $\frac{1}{16}$ the distance, as MR. Space along the line by moving the dividers clockwise and counterclockwise to positions 2, 3, 4, and 5. If MR was taken too large, as shown by LF, adjust the setting and respace the line until the correct results are finally obtained.

The setting of the dividers after each trial is adjusted by slightly turning the hair-spring adjustment screw until the correct length is obtained. For extremely fine divisions, bow dividers are more convenient.

Hints on Penciling
The pencil should always be *drawn* not *pushed*. Lines are generally drawn from left to right and from the bottom to the top. By keeping a drawing in a neat, clean condition when penciling, the use of the eraser on the finished drawing will be greatly diminished.

Dimension Drawings
Every dimension necessary for the execution of the work indicated by the drawing should be clearly stated by figures on the drawing. This is so that no measurements need to be taken in the shop by scale. All measurements should be given with reference to the base or starting point from which the work is laid out and also with reference to center lines.

There are two ways to dimension a drawing: the *aligned system of dimensioning* and the *unidirectional system of dimensioning* (see Figure 1-35). When using the aligned system, all dimensions are placed in line with the dimension line. If the unidirectional system is used, all dimensions are read from the bottom of the drawing.

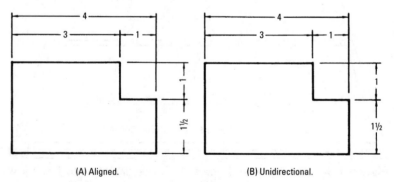

(A) Aligned. (B) Unidirectional.

Figure 1-35 The system of dimensioning.

There are certain types of lines used in dimensioning a drawing. The four basic lines are extension, dimension, centerlines, and leaders (see Figure 1-36). Extension lines are extensions of the object lines. They start $\frac{1}{16}$ inch past the object and extend $\frac{1}{8}$ inch past the last dimension line. Dimension lines are placed between two

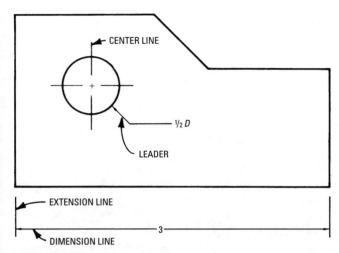

Figure 1-36 Basic dimension lines.

extension lines and are terminated by arrowheads. In mechanical drawing, the dimension line is broken in the middle and the dimension is placed in the gap. In architectural drawing, the dimension line is solid and the dimension is placed on top of the line.

Centerlines are composed of long and short dashes and are used to locate the center of circles and cylindrical shaped objects. A leader is an inclined line that has a horizontal shoulder. The inclined line is terminated by an arrowhead and shows the location of a special feature.

All dimensions that a builder may require should be put on a drawing so that no calculations are required by the builder. For example, it is not enough to give the lengths of the different parts of the object; the overall length (which is the sum of all these lengths) should be placed outside, in which case an arrow should be put in to indicate the proper position of the figures.

Electronic Drawing

CAD is generally considered to mean computer-aided (automated) drafting. A few texts use it to mean computer-aided design. Like the drafting tools it replaces, the CAD system is a graphics system.

A key concept is that a CAD system is based on a computer. It harnesses the computational facilities of the computer to automate many design and drafting tasks and, thus, to aid the designer.

Computers

By definition, a computer is necessary for CAD. Personal computers are becoming more common, especially in business (see Figure 1-37 and Figure 1-38). Many builders and even self-employed carpenters are using a personal computer for estimating, accounting, and letter writing.

Figure I-37 A personal computer can be used for electronic drawing.
(Courtesy American Standard, Inc.)

A personal computer is an electronic device that takes in data, processes it, and gives out information based on that data. For example, a contractor with computerized accounting types all the charges for materials and labor for various projects. The computer processes that data and prints out bills ready to mail, prints out paychecks for employees, and shows the contractor which jobs are behind schedule and which are over budget. The computer processes the raw data into a form of information that is more useful.

A computer system consists of hardware and software. *Hardware* is the physical device that you can see and touch (see Figure 1-39). The *keyboard* is like a typewriter keyboard that sends signals representing letters and numbers to the processor. The *central processing unit* is made of electronic components and circuit boards that sort

Figure 1-38   Tools used for mechanical drawing.
(Courtesy American Standard, Inc.)

and analyze the data. The *memory* is also made of electronic components and circuit boards, but it just stores data for use by the processor. Data in the memory is organized into groups called *data files*. The *screen* (or *video display*) is like a television screen. It takes information such as letters, numbers, or graphics from the processor and displays it for you to look at. The *printer* takes information from the processor and puts it down on paper.

Software is a list of instructions stored in the memory. You can't see it or touch it because it only exists as electronic impulses within the hardware. Each instruction is performed in order by the hardware. Software controls what the hardware does and controls the

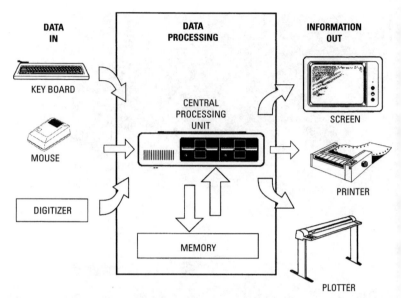

Figure 1-39 Computer system hardware. The three basic functions of a computer system are data input, data processing, and information output. Arrows indicate the flow of data and information.

flow of data through the whole system. A list of instructions is called a *software program*. The program makes the hardware perform specific operations that can help you keep track of your accounts, write a letter, or play a video game.

Personal computers are very versatile. They can also be used as the basis for a CAD system.

In conventional or manual drafting, a drawing is created by making marks on a sheet of paper. The drawing is modified by erasing marks and making new ones. At any time, the drawing can be viewed with the unaided eye.

A CAD system is used to create and maintain drawings in electronic form. Whereas the drawing itself exists in the computer's memory, there is no physical drawing until one is printed or plotted on paper.

The CAD drawing is a list of objects stored in the computer's memory—an electronic list of lines, circles, arcs, points, and so on. As such, it is not accessible like a traditional drawing. However, with the aid of the CAD software, these electronic drawings can be displayed on the computer's screen or drawn on paper.

The CAD system is a system of hardware and software components. The system is used to create, maintain, and display these electronic drawings. The basic requirement for the computer is that it includes graphics capabilities, as CAD is graphics intensive.

Why Use CAD?

Virtually every prospective CAD user must deal with this question. Those who are not convinced of the value of CAD ask the question, and those who are convinced are forced to answer it. Conventional drafting techniques work, so why change?

Usually, the first issue that arises is that of increasing drafting productivity. Many CAD system vendors quote average productivity increases on the order of 3 to 1, higher increases for certain activities. Skeptics, on the other hand, typically reply that they can manually draw faster than CAD operators can with the CAD system.

There is no clear-cut answer to the productivity question. There are fast and slow CAD operators, just as there are fast and slow manual drafters. In fact, CAD is neither slower nor faster. The CAD system is only a tool. It is the operator that determines the speed of the drafting operation, just as a carpenter determines the speed at which a power saw cuts wood.

Most CAD users are able to produce drawings faster with CAD systems than with manual methods after only a short period of time.

Even if CAD did not allow drafters to produce drawings faster than they could manually, there would still be many benefits. One is accuracy. A CAD system provides an arbitrary level of accuracy in the drawing database—accuracy that is independent of the hardware used to create the drawing. Manual drafting methods do not allow arbitrary accuracy. The width of a pencil line limits the accuracy with which a line can be drawn or measured afterward. Dimensional notations can be arbitrarily accurate, but the linework cannot. In a CAD drawing, the geometric elements can be essentially as accurate as the dimensional notations.

The accuracy of a CAD drawing becomes significant in connection with another benefit: reuse of data. Because the geometric data in a CAD drawing is stored electronically, it can easily be transferred to other drawings (in whole or part) with no loss in quality or accuracy. By contrast, transferring data from one conventional drawing to another is only a slight improvement over creating the second drawing from scratch. There are photographic techniques that can be used to eliminate time-consuming tracing, but most of these techniques involve some sacrifice in quality and accuracy.

CAD offers the possibility of creating a living database. Consider the process of constructing buildings. Using conventional methods, an architect designs a building and records the design as a set of drawings. In the process of designing the building, he or she makes material take-offs and cost estimates from the drawings. The contractors bidding on the project must also make material take-offs and cost estimates. However, material quantities are not part of the drawing, so the contractors repeat the process from scratch.

The successful contractor uses the design drawings as a set of instructions for the construction of the building. However, since architectural drawings are not detailed exhaustively, there are changes that occur in the configuration of the building, selection of materials, and other aspects of the building. Consequently, the design drawings seldom if ever show the building as it is finally constructed. Yet the owner needs an accurate set of drawings to properly maintain the building. It is common practice to annotate a set of drawings with the changes made during construction, and to turn that set over to the owner when the building is complete. In reality, neither the design drawings nor the as-built drawings are truly accurate. In addition, even if they were, design and construction drawings are not very well suited to facilities management. Consequently, the building owner must make a building management database from scratch.

CAD offers an alternative. CAD drawings can easily be structured to include material quantities, relieving the contractor of the need to make material take-offs from scratch. They can also be structured so that the contractor's shop drawing submittals can easily be incorporated if submitted in CAD form. After shop drawings are merged into the design drawings, the CAD drawing can be an accurate portrayal of the building's configuration. If the as-built CAD drawings are turned over to the owner in electronic form, he or she can use the drawing database as the basis of the facilities management database. Finally, if the facilities management system includes a CAD system, drawing updates can be incorporated into the facilities management cycle. As the building configuration changes through maintenance and remodeling, the drawing database can change accordingly.

The CAD System

The basic personal computer with a graphics display system will host a CAD system. Such a minimal system would not be the most effective, but it would provide the basic capabilities.

A graphic input device (also called a *pointing device*) is a necessity in a CAD system. It is used to position a cursor or pointer on the graphics screen and to indicate menu selections and graphic data points. The PC keyboard includes arrow keys that can be used as the graphic input device in the bare-bones system. There are better alternatives, however.

The *mouse* is much easier to use and is inexpensive. The disadvantage of using a mouse is that it can only be used to point to areas of the screen. It cannot be used with tablet menus, as can *digitizer tablets*.

The digitizer tablet performs the same basic task as the mouse, although with a different mechanism. It is an improvement over the mouse, because it can be used to point to areas on the screen and to make selections from a tablet menu. In addition, a digitizer can be used to trace a manually prepared drawing, which loads the drawing into the computer so that it can be stored and edited electronically.

Both mouse and digitizer provide a way to accomplish the hand-eye coordination of manual drafting on the computer. As the pointing device is moved over a physical surface (a desktop or the digitizer tablet), a *cursor* (or spot of light) moves on the screen.

With the graphics-capable computer and pointing device, the CAD operator can create CAD drawings. As with any computer program, however, memory space may limit the size of the drawing data file. With CAD, the data file is the CAD drawing, so high-capacity memory may be required for large drawings.

Most applications require an output device capable of printing or plotting the drawings on paper. For small format drawings, where high quality is not required, a *dot-matrix graphics printer* may suffice. Even though the prints of the drawings may not be impressive, the CAD drawing itself is exact.

A *pen plotter* is a practical requirement for the typical CAD user. The pen plotter produces a physical reproduction of the CAD drawing on paper. Again, the plotter cannot match the original electronic version of the drawing, but it can far surpass the manual drafter, typically producing plots accurate to a few thousandths of an inch.

The last key component of the CAD system is the CAD program. Actually, most CAD software packages consist of several programs operating in concert, but the distinction between a program and a system of programs is not important for our purposes.

CAD Concepts

With the components of a CAD system established, the concepts behind the system can now be explored.

Drawing Primitives

Drawings are symbolic representations of real-world objects. No drawing, however detailed, can be confused with the real thing. In fact, architectural drawings are not intended to portray a realistic image. Instead, they are intended to convey certain types of information in the most concise manner possible.

Although *architectural drawings* do not depict objects in a realistic manner, they do convey certain types of information much better than a pictorial drawing. A *scale drawing* conveys dimensional information about the real object represented in the drawing through dimensional notations and through the relative sizes of the drawing symbols. Simple patterns of lines can be added to indicate materials, assembly methods, and finish. Drawings consisting of several views allow the experienced observer to construct a more accurate mental picture of the real object than could be conveyed by a pictorial drawing.

The complex symbols of a drawing can be reduced to simpler symbols—familiar geometric entities. Any drawing is actually a collection of points, lines, circles, and arcs (see Figure 1-40). In fact, a drafter creates piecemeal the complex symbols that the observer perceives in a drawing.

These simple elements of a drawing are called *primitives*. CAD systems are not limited to such a small set of primitives. A CAD system includes not only the four primitives listed but also rectangles, regular polygons, ellipses, complex curves, and text. This larger set

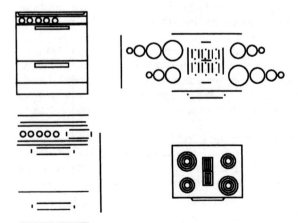

Figure 1-40 A complex symbol consists of simpler objects called drawing primitives. *(Courtesy Generic Software)*

of primitives corresponds closely to the set of symbols and tools used by drafters—circle and ellipse templates, French curves, and so on.

The Drawing Database

A *database* is a collection of data organized to facilitate storage, search, and retrieval. A telephone book, which is a list of names, addresses, and telephone numbers, is a database. Given a name, you can find the telephone number associated with it.

A conventional drawing on paper can be viewed as a collection of primitives, as noted previously. By making a suitable list of the primitives in one person's drawing, it should be possible for another person to reconstruct the drawing from the list. Simply listing the primitives is not enough, of course. The list would have to include the name of each primitive, its location and orientation on the drawing, and geometric properties. Indeed, to allow an exact reproduction, the list would have to include such items as line width and the force with which the pencil was pressed against the paper. If such a list could be constructed, it would be a database.

A CAD drawing file is just such a list. The list contains a description of each primitive along with the data required for the CAD program to draw the primitive on the screen, on a printer, or on a plotter (see Figure 1-41).

DATABASE TEXT FILE

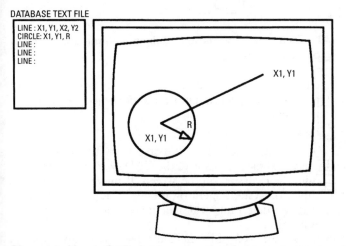

Figure 1-41 A CAD database is a list of objects in the drawing. Each item in the list contains the parameters required to construct an image of an object at any size on the screen or with the plotter on paper.

(Courtesy Generic Software)

It is not essential for the CAD operator to know the exact contents of the drawing file, but it is helpful to have a general understanding of how primitives are stored there.

To draw a line, a manual drafter must know the locations of the two endpoints. The CAD system is no different. For each line primitive in the database, the description must include the locations of the two endpoints. For each circle, the description must include the location of the center point and a point on the radius (see Figure 1-42).

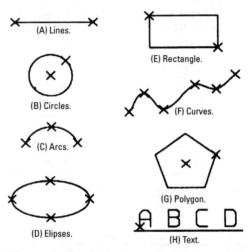

(A) Lines.

(E) Rectangle.

(B) Circles.

(F) Curves.

(C) Arcs.

(D) Elipses.

(G) Polygon.

(H) Text.

Figure I-42 Each type of primitive is defined with the use of one or more date points. *(Courtesy Generic Software)*

Drawing an object adds a description of an object to the list. To erase an object, the CAD program must find the object's description in the list and delete it.

Pointing and Picking

Since a CAD drawing is electronic, the objects in it cannot be manipulated by physical means. Instead, the CAD system maintains an electronic pointer (or cursor) on the screen. It moves the cursor as directed by the CAD system operator, it keeps track of the cursor's location, and it can determine whether the cursor is located on or near a drawing object.

Moving the cursor to a point on the screen or to an object in the drawing is called *pointing*. It is a key method of providing graphic data to the CAD system. The CAD system moves the cursor in response to signals from a pointing device, a piece of hardware that can be physically moved. Typical pointing devices are mice and digitizer tablets.

The CAD system continuously monitors the pointing device and positions the cursor on the screen accordingly, but it does not change the drawing as it moves the cursor. The cursor does not exist in the drawing but is superimposed *on* the drawing, just as a manual drafter moves a parallel bar and triangle over the drawing board. As the cursor moves across the screen, the CAD system displays numbers at the top of the screen. The numbers show the size of the object you are drawing in feet and inches.

Graphic data points are actually entered into the drawing by *picking* them. A pointing device has a pick button that is pressed to signal the CAD system to use the set of coordinate values indicated by the cursor.

An object in the drawing can be brought to the CAD system's attention by placing the cursor over it and pressing the pick button. It takes note of the coordinates picked and searches the drawing database for an object that falls on the point.

Commands

A CAD system waits patiently for directions, a command. A command is entered by typing on the keyboard. The system executes the command, modifying the drawing database and updating the screen display as required. Then it waits for another command.

Every CAD system has a set of commands it recognizes and a method of punctuating those commands. This is called a command *syntax*. Some large systems have very complex commands, resembling English sentences. Others have simpler one-word commands but still require punctuation at the end of the word—a space or carriage return.

In the examples presented here, two-letter commands are used.

A Simple Drawing

The purpose of this section is to introduce you to the basics of drawing with CAD. The subject will be a modest lakeside cabin.

The cabin will be 24 feet × 32 feet, with a floor area of 768 square feet. There will be one large room, a bathroom, and a kitchen. Before we start, though, let's take a few moments to make a rough sketch on paper of the cabin's floor plan. This will give us a foundation to work on (see Figure 1-43).

Draw the Outside Walls

To draw the outside 4-inch walls we will use two rectangles, one inside the other (see Figure 1-44).

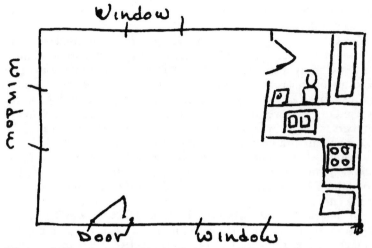

Figure 1-43 A pencil sketch of the cabin's floor plan is a useful preliminary step in the drafting process. *(Courtesy Generic Software)*

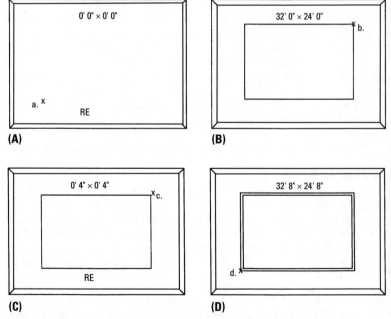

Figure 1-44 Rectangles form the outside walls.

Commands	Instructions
RE (Rectangle)	Move the cursor to the bottom-left corner. Pick a starting point *(Aa)*. Move the cursor toward the upper-right corner and pick the second point at 32 feet × 24 feet *(Bb)*. As soon as you pick the second point the CAD system displays the rectangle on the screen.
RE (Rectangle)	The current 0,0 point is the upper-right corner. Move the cursor toward the upper-right and pick the first point as 4 inches × 4 inches *(Cc)*. This point now becomes the origin (0,0). Move the cursor toward the bottom-left corner. Pick the second point at 32 feet 8 inches × 24 feet 8 inches (8 inches is used because we added 4 inches to the other side) *(Dd)*. The CAD system displays the second rectangle on the screen.

Now that we have the first floor of our cabin and a 4-inch wall, let's draw the interior walls.

Add Interior Walls
We will place the bathroom in the upper-right corner of our cabin. The dimensions are 8 inches × 5 inches (see Figure 1-45).

Add Kitchen Counter
Next we will draw an outline of the counter (see Figure 1-46).

Now that we have the first floor of our cabin and a 4-inch wall, let's draw some of the furnishings. The furnishings are a bathtub, toilet, bathroom sink, stove, refrigerator, window, door, and kitchen sink. There could be many more, but you can tend to the details later.

Make Window
Because we want to draw each of these fixtures only once, we are going to create them as components (sometimes referred to as *symbols*).

A *component* is a number of primitive objects grouped together and then manipulated as a single object.

The ability to define, place, and manipulate symbols as single objects is one of the features that gives CAD designers a significant advantage over their conventional counterparts. The use of drawing symbols or components speeds the drawing process in two ways. First, only one command needs to be entered to draw a complex

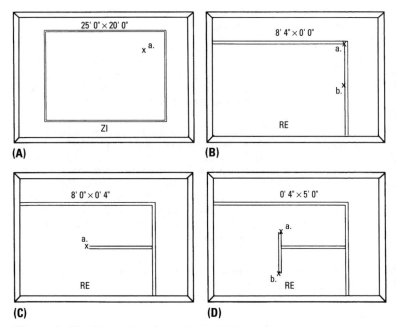

Figure 1-45 Rectangles form the interior walls.

Commands	Instructions
ZI (Zoom In)	Move the cursor to the center of the bathroom area in the upper right and pick a point *(Aa)*. The CAD System displays a close-up view of the bathroom area *(B)*.
NP(Snap Near Point)	Move the cursor near the inside upper-right corner and snap to the corner *(Ba)*. This sets the origin at that point.
RE (Rectangle)	We want the wall 4 inches thick, so move the cursor down to 5 feet 4 inches and pick the starting point on the inside wall *(Bb)*. Move the cursor left and up. Pick the second point at 8 feet × 4 inches *(Ca)*.
RE (Rectangle)	Move the cursor up 2 feet and pick the starting point *(Da)*. Move the cursor down and pick the second point at 4 inches × 5 feet *(Db)*. (The other side of the doorway is done with a third rectangle. After just a few commands the bathroom is framed in.)

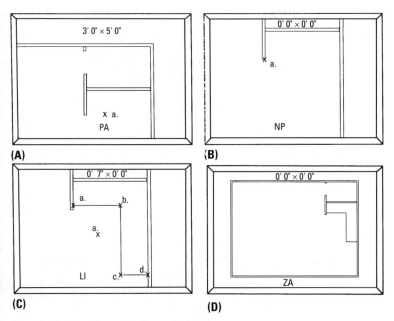

Figure 1-46 Draw outline of kitchen counter.

Commands	Instructions
PA (Pan)	Pick a point in the kitchen area *(Aa)*. (The CAD system shifts the view downward *(B)*.)
NP (Snap Near Point)	Pick the lower-right corner of the kitchen wall *(Da)*.
LI (Line)	LI is the LINE command. Move the cursor up 2 inches and pick the starting point *(Ca)*. Move the cursor right 5 feet 5 inches and pick the second point *(Cb)*. Move the cursor down 7 feet 4 inches and pick the third point *(Cc)*. Move the cursor right 2 feet 7 inches and pick the last point, joining the outside wall *(Cd)*.
ZA (Zoom All)	To view all of the drawing *(D)*.
DS (Drawing Save)	Save the drawing. The CAD system puts the drawing file into the computer's memory.

object that could otherwise take scores of commands. Second, the designer can think in terms of placing symbols rather than constructing them from primitive objects.

To create a component, first draw the symbol with primitives and then convert it into a component (see Figure 1-47).

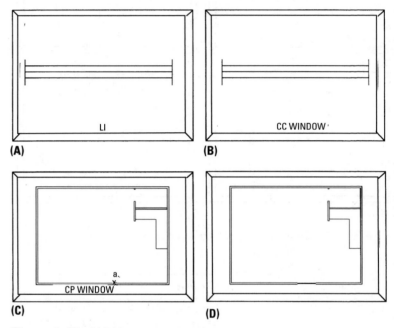

Figure 1-47 Window component.

Commands	Instructions
LI (Line)	Draw a window symbol with 4 lines (*A*).
CC (Component Create)	Create the component with the name WINDOW (*B*).
CS (Component Save)	Save the component in memory as a file named WINDOW.

Now that the window component is created it can be placed in the drawing in a single operation (see Figure 1-48).

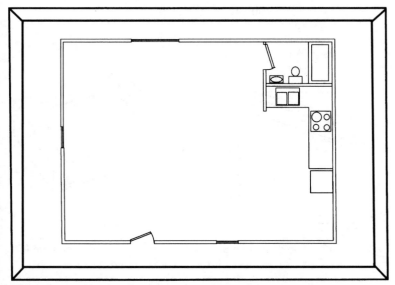

Figure 1-48   Components for other windows and furnishings are created and placed in the same way. *(Courtesy Generic Software)*

Commands	Instructions
DL (Drawing Load)	First, load the cabin drawing into the CAD system from memory.
CP (Component Place)	Type the name WINDOW. Pick a point on the bottom line (outside wall) *(Ca)*. The window is placed within the wall *(D)*. The CAD system automatically places the window symbol within the wall.

Now that the window component is created it can be placed in the drawing in a single operation (see Figure 1-48).

Plotting

To get the drawing on paper it must be plotted or printed.

Many new and prospective CAD users confuse the plot of a draw-ing with the drawing itself. And, for some users, there is no harm done, as they consider the CAD system strictly as an electronic

PLOTTERS

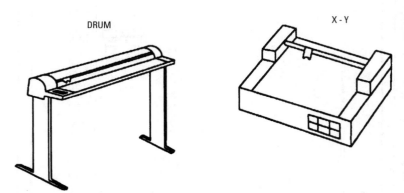

Figure 1-49 The drum pen plotter is the most common medium format plotter. Small format X-Y plotters can be placed on a desktop.
(Courtesy Generic Software)

replacement for the drawing board. However, the CAD system is far more powerful than a drawing board, and it is necessary to mentally separate the drawing from a plot of the drawing.

Even so, plots are essential. It is far easier to carry a roll of prints to the construction site, the building department, or the customer's office than to carry a computer.

The most commonly used small to medium plotter is the *drum pen plotter* (see Figure 1-49). In this type of plotter, the paper is moved back and forth under the pen along one axis, and the pen is moved back and forth along the other axis. By varying the relative speeds of the two movements, the plotter can produce straight lines at any angle, and complex curves. Drum pen plotters are compact and relatively inexpensive.

Summary

Drawing is a tool carpenters and builders can use to formalize their own thoughts, document projects, and communicate with others. Refining drawings through a series of drafts is the way to develop meaningful drawings.

Mechanical drawing is done with the aid of drawing instruments. Using the best quality instruments makes good drawing easier. Small instruments (such as pencils and triangles), as well as larger equipment (such as a drawing board or drafting machine), are used to set ideas down on paper. Special techniques from pencil sharpening to handling the instruments must be used with frequent practice to develop proficient skills.

Electronic drawing uses a computer-aided drafting (CAD) system to improve efficiency and accuracy in drawing. The CAD system is based on a computer with special devices to enter and print out drawings. Software in CAD systems controls the hardware devices to make drawing more accurate and efficient.

Review Questions

1. What is the difference between a compass and a divider?
2. What is a T-square? How is it used?
3. What is a protractor used to measure?
4. What is a drafting machine? Why would you want one?
5. How does CAD help a builder or contractor cooperate with an architect and owner?
6. What function does a mouse or digitizer play in a CAD system?
7. How do you get a drawing out of the CAD system so you can use it?
8. The term *mechanical drawing* means drawing with the aid of _____ instruments.
9. Which pencil is used for lettering?
10. What instruments are used for drawing straight lines?

Chapter 2

Safety

There are two reasons for having safe work habits while performing duties as a carpenter: one is the health and well-being of the individual, and the other is financial. Accidents are prevalent in the building trades. They cause partial or total disability in many cases and make safe working practices well worth the time and effort.

Safety is knowledge-based, as well as part skill and a lot of proper attitude. Carpenters must develop an attitude early in their careers that allows for the safe means of doing a job. Taking time to study how to do something safely is well worth the time and effort. Learning the proper use of equipment, proper techniques, and proper procedures will go a long way in maintaining safe work habits and a safe workplace (see Figure 2-1).

Figure 2-1 Use proper equipment, setups, and procedures to do the job safely.

Pain and suffering are enough to cause anyone to think twice. However, most of us do not think in terms of pain and suffering happening to us. We always think it will be someone else—it can't happen to me! Look around you and see how many people you can talk with who have been in an on-the-job accident.

If pain and suffering are not enough, think of the financial problems caused by not having an income and having to pay hospital and doctor bills. The safety of people within an area is, of course, the responsibility of the individual in charge of that area, whether it is a private home, a public building, an office or a factory, or a construction site. Of course, no one wants anyone else to be injured, but the legal obligations of a contractor or owner of property extend beyond humanitarian concerns.

Some safety rules applying to the carpenter when on the job deal with topics such as clothing to be worn while working, protective equipment, hand tools, power tools, good housekeeping, decks and floors, excavations, scaffolds and ladders, lifting and carrying, falling objects, fire protection, and, of course, first-aid.

Clothing

Trousers and overalls should fit properly. Trousers should have legs without cuffs. Shirts and jackets should be kept buttoned. Sleeves should also be buttoned or rolled up. Never wear loose or ragged clothing. Keep your clothing in good repair.

Do not wear tennis shoes or sneakers on the job, except when roofing calls for it. Wear heavy rubber-soled shoes with steel toes for protection of the feet (see Figure 2-2).

Figure 2-2 Use proper clothing for the job.

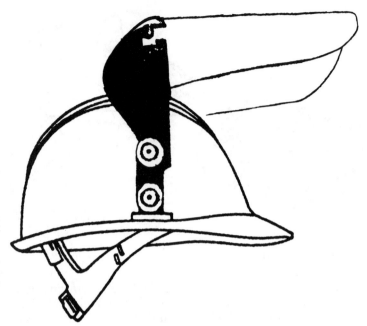

Figure 2-3   Hard hat with flip-down shield.

Wear head protection. If there is danger of falling objects, wear a hard hat (see Figure 2-3). Most carpenters use a cap or hat to protect their eyes from the sun. In addition, safety goggles or visors may be required when working with materials or situations more hazardous to the eyes (see Figure 2-4).

Protective Equipment

Hard hats, steel-toed shoes, eye protection, and gloves should meet standards specified by the Occupational Safety and Health Administration (OSHA) and other agencies. If working around dusty areas, wear a respirator with the proper filters to protect your lungs. Gloves and breathing protection should be used when cutting or handling treated wood products.

Hand Tools

Keep your tools sharp and use the proper tool for the job. Dull tools cause injury and damage. Keep the handles tight. Oil those that require such attention. Clean tools that are oily or greasy. Be careful when using your fingers or hand as a guide for starting a cut.

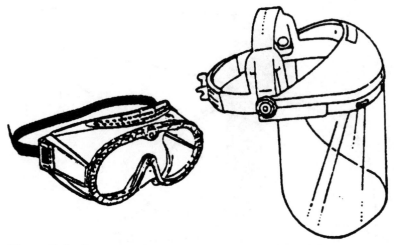

Figure 2-4 Eye protection.

Hold tools correctly. Also, take care in handling or carrying tools between jobs. Whenever tools are being carried from place to place, point edges away from the body or turn downward. Store tools in chests or tool boxes when not in use.

Power Tools

Working with power tools demands that you know what the tool does and how it does it. Working with the tool safely is of utmost importance. Remember that some quick shortcuts can be hazardous. Power tools can eliminate hours of construction time, but they can cost a lifetime of pain and misery if not properly handled. Wear eye protection when using any power tool (see Figure 2-4).

Circular Saw Safety

Following are some safety tips when working with circular saws:

- Support the stock being cut in such a way that the groove made by the saw blade does not close up and bind during the cut or at the end of the cut.
- Wear eye protection.
- Clamp small pieces to a bench or sawhorse. Cut thin materials only with adequate support.
- Do not cut the sawhorse, bench, or supporting device.
- Adjust the blade so it cuts with no more than $1/8$ inch showing through the stock being cut.

- Check all adjustments on the saw guide to make sure they are tight.
- Make sure the outlet used for the saw is grounded adequately and according to the *National Electrical Code*'s requirements for temporary power sources.
- Make sure the saw base is on the stock with the blade clear before turning on the switch.
- Never reach under the material being cut.
- Stand to one side when cutting.
- Keep both hands on the saw when power is on.
- Never change blades unless the power cord is disconnected from the electrical outlet.
- Always use a sharp blade.

Table Saw Safety

Some construction sites use table saws to do woodworking for cabinets and trim. Chapter 7 provides more detailed information on the table saw. [You may also want to consult *Audel Carpenters and Builders Millwork, Power Tools, and Painting: All New Seventh Edition* (Wiley Publishing, Inc., 2005) for more information.] Here we mention some of the most important aspects of table saw safety:

- Extend the blade no more than $1/4$ inch above the stock being cut.
- Keep your hands a safe distance (at least 4 inches) from the saw blade.
- Lower the blade and turn off the power when finished.
- Make sure the blade is sharp.
- One edge and one surface should be jointed before being placed on the saw bed for cutting.
- Stand to one side when cutting.
- Use a fence or miter gage for guiding the stock.
- Use the right blade for the job.
- Wear eye protection.

Radial Arm Saw Safety

Most construction sites have a radial arm saw handy for quick cuts of rough lumber to given lengths. This saw can speed up the job, but it can also be very dangerous when improperly operated. Following are some of the safety tips to keep in mind when using this type of saw:

- Wear eye protection (see Figure 2-4).
- Hold stock firmly on the table and against the fence for cross-cutting operations. Support long boards so they are level with the table.
- Check to see if the saw is slightly off-balance toward the back so the blade has a tendency to move back toward the fence when turned off. Check clamps and stability of the stand.
- Make sure the guard and kickback device are in position.
- Return the saw to the back of the table when the cut is finished. Don't remove stock from the table until the saw is moved back and the blade has stopped turning.
- When cutting, keep your hands at least 6 inches from the blade at all times.
- Make no adjustments unless the blade has been stopped.
- Keep the table clean and free of scraps.
- Make sure the area around the saw is free of scraps that could cause a person to trip and fall into the saw blade when it is turning.
- Never attempt to clean the tabletop when the blade is running.
- Always pull the blade toward you when crosscutting.
- Always feed stock into the blade when ripping. The bottom teeth should be turning toward you.

The radial arm saw is a handy tool to have for cutting rafters, but be especially careful when cutting compound miters such as cheek cuts.

Portable Electrical Drill Safety

Even simple tools such as drills can be the source of injury on the job if not properly handled. Following are some good safety rules to keep in mind:

- Wear eye protection (see Figure 2-4).
- Select the right bit for the job. Mount it in the chuck tightly.
- Do not allow the stock being drilled to move while you are working on it.
- Make sure the drill is *off* when you plug it in.

- Check to see if the bit is properly mounted by turning the drill on and checking for proper operation.
- Place the drill bit into the punched hole before turning on the drill.
- Hold the drill firmly and at the desired angle with both hands if necessary.
- Keep the drill properly aligned with the hole while drilling to avoid binding and bit breakage.
- Withdraw the bit several times when drilling a deep hole to remove the cuttings.
- Remove the bit from the drill when you are finished with the job. Properly store the drill so the cord does not cause an accident.

Other Power Tools

From time to time the carpenter may also be called on to use other power tools, such as the jointer, sander, jigsaw or band saw. The portable router is also a handy power tool. Each of these tools should be handled with care and manufacturer's operation instructions should be followed when they are being used on the job.

Good Housekeeping

Just as at home, this term refers to keeping the job site clean and organized. Neatness creates a safe workplace and an efficient one. Organize building materials so they are easily and safely reached. Don't allow nails, bolts, empty cans, bottles, wire, or anything to accumulate to cause someone to trip or fall. A good appearance is contagious—it aids in the mental attitude toward the job and the worksite. Better efficiency results in fewer accidents and fewer lost work hours.

Excavations

Sides of excavations should be sloped to allow for safe removal of materials and to prevent cave-ins. Never jump into an open trench unless it has been reinforced properly against cave-in. Check the shoring to make sure it has not cracked or shifted. This is very important after a rain.

Scaffolds and Ladders

Scaffolds should be checked to be sure they are safe to handle a load four times greater than the normally expected load. Use experienced people to construct scaffolding. Ladders should be clean and inspected for cracks or loose screws or bolts (see Figure 2-5). Wooden ladders should not be painted in order that they can be checked often for cracks or safety defects. There are many safety rules to be followed when working on ladders and scaffolding. [See *Audel Carpenters and Builders Layout, Foundation, and Framing: All New Seventh Edition* (Wiley Publishing, Inc., 2005) for more information.].

Figure 2-5 Unsafe site-built ladder.

Decks and Floors

Carpenters should not work on floors or decks that are not firm or solid, especially when using hand or power tools. The surface should be smooth and not slippery. Install guardrails when floor openings are in heavily traveled areas. In bad weather, keep the ice and snow from surfaces where work is being done.

Falling Objects

When working on the top floor or on the roof, keep in mind those below. Anything dropped or thrown over the side during the time you are doing a particular job may injure someone.

Don't place tools on ladder steps or the edge of scaffolding. Windowsills and other surfaces may be a handy place to leave the hammer or pliers or some other tool for a minute, but the tool may fall or be knocked off and hit someone else, or in some cases, the person who put it there. Keep away from materials being hoisted. Keep away from areas where something may be dropped on you as you walk by. Wear a hard hat whenever you are in an area where there might be falling objects.

Lifting and Carrying

Lifting of heavy objects, if not done properly, can cause hernias, back problems, and strained muscles. When lifting heavy objects, there is a right way and a wrong way. The wrong way is to bend over from the waist and lift; the correct way is to bend the knees and pick up while straightening the knees. Keep the back as close to vertical as possible, and lift with the legs rather than the back. Also, put down heavy loads by bending the knees and keeping the back straight. Do not attempt to lift loads that cause muscle straining.

Don't twist your body or make shifting movements with your feet when carrying a heavy load. Get someone to help when carrying long pieces of lumber.

First-Aid

Check with the Red Cross to obtain instruction in first-aid. At least one person on the job site should have had first-aid training. Another thing to keep in mind is the location of the local hospital and how to get there fast. Check to make sure their emergency number is handy on the site.

It is a good idea to check with the medical facilities in the area where you are working to find someone who specializes in *hand reconstructive surgery.* Saw accidents on the job have a tendency to do a great deal of damage to fingers and hands. They can be saved if properly and quickly treated. Call the hospital to see if a hand specialist is on the staff and what to do in case of an accident on the job site.

Summary

There are two reasons for having safe work habits while performing duties as a carpenter: one is the health and well-being of the individual, and the other is financial.

Safety is knowledge-based, as well as part skill and a lot of proper attitude. Pain and suffering are enough to cause anyone to think twice.

Safety rules that apply to the carpenter when on the job deal with topics such as clothing to be worn while working, protective equipment, hand tools, power tools, good housekeeping, decks and floors, excavations, scaffolds and ladders, lifting and carrying, falling objects, fire protection and, of course, first-aid.

Review Questions

1. What are the two reasons for having safe work habits?
2. How does proper clothing add up to safe working conditions?
3. What and who is OSHA?
4. How does having a sharp tool to work with make working safer?
5. Why should you wear eye protection while using the hammer or saw?
6. How can good housekeeping aid in making safe working conditions?
7. Why do you need guard rails anytime the project work area is more than 6 feet off the ground?
8. If working over 25 feet above the ground or water what is needed to protect those who might fall?
9. Where do you obtain instructions in how to administer first-aid?
10. How can ladders and scaffolds contribute to unsafe working conditions?

Chapter 3

Guiding and Testing

In good carpentry, much depends on accuracy in measurement and in fitting parts together at the required angle. To ensure this accuracy, various tools of guidance and direction are used, otherwise joints and so on could not be made with precision.

Straightedge

This tool is used to guide a pencil or scriber when marking a straight line and when testing a faced surface (such as the edge of a board) to determine if it is straight. Anything having an edge known to be straight (such as the edge of a steel square) may be used. However, a regular straightedge is preferable.

The straightedge may be made of wood or steel, and its length may be from a few inches to several feet. For ordinary work, a carpenter can make a sufficiently accurate straightedge from a strip of good straight-grained wood (see Figure 3-1), but for precision work, a steel straightedge should be used. Wood is not recommended for precision work because of its tendency to warp or spring out of shape.

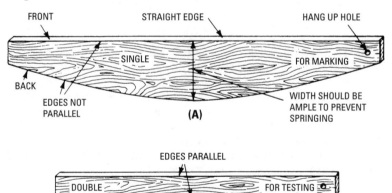

Figure 3-1 Wooden straightedges. When well made, they are sufficiently accurate for ordinary use. (A) Single straightedge; (B) double straightedge.

Figure 3-2 shows the correct and incorrect methods of holding a straightedge as a guiding tool, and Figure 3-3 shows how and how not to hold the pencil when marking stock.

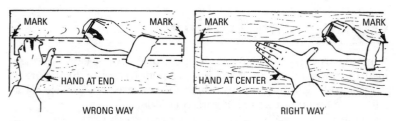

Figure 3-2 The incorrect and correct methods of using a straightedge as a guiding tool. To properly secure the straightedge, the hand should press firmly on the tool at its center, with the thumb and other fingers stretched wide apart. Otherwise, the straightedge may slip.

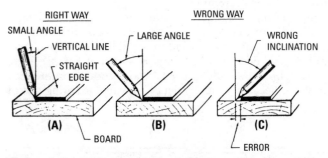

Figure 3-3 Right and wrong inclinations of the pencil in marking with the straightedge. The pencil should not be inclined from the vertical more than is necessary to bring the pencil lead in contact with the guiding surface of the straightedge (A). When the pencil is inclined more and pressed firmly, considerable pressure is brought against the straightedge, tending to push it out of position (B). If the inclination is in the opposite direction, the lead recedes from the guiding surface, thus introducing an error that is magnified when a wooden straightedge is used because of the greater thickness of the straightedge (C).

The basic skill of testing the straightness of lumber is just as important today as in 1923 when these figures first appeared in Audel (see Figure 3-4 and Figure 3-5).

Square

This tool is a 90° (or right angle) standard and is used for marking or testing work. There are several common types of squares (see Figure 3-6):

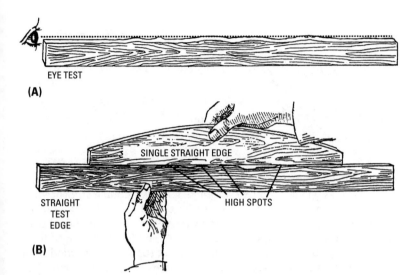

Figure 3-4 Eye and single straightedge tests for longitudinal uneven-ness of surface. When planing the edge of a board, after a few strokes of the plane, hold the board toward the light, close one eye, and sight along the board as at (A) to see if the surface is plane or uneven. With practice, a good straight surface can be obtained. For precision used in addition to the eye, test the straightedge as at (B). Hold board with straightedge between the eye and a bright light and any unevenness will show plainly.

- Try square
- Miter square
- Combined try-and-miter square
- Framing or steel square
- Combination square

Try Square
In England, this is called the *trying square*, but here it is simply the *try square*. It is so called probably because of its frequent use as a testing tool when squaring up mill-planed stock. The ordinary try square used by carpenters consists of a steel blade set at right angles to the inside face of the stock in which it is held. The stock is made of some type of hardwood and is faced with brass to preserve the wood from damage.

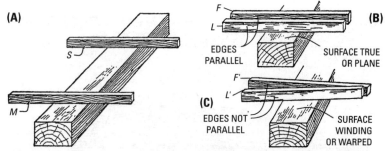

Figure 3-5 Double straightedge test for transverse unevenness of surface known as winding. To test for wind, take two double straightedges and place them across the surface to be tested as at (A). Now if the surface has no wind (that is, if it is at right angles to the side of the board at the sections tested), then the edges of M and S, when viewed from the end of the board, will be parallel as LF at (B). If there is wind (that is, if the sections tested are not at 90° to the side of the board), then the edges of M and S will be inclined to each other when viewed at the end as L', F', at (C).

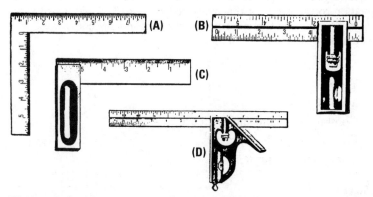

Figure 3-6 Various types of squares: (A) represents a steel square; (B) a double try square; (C) a try square; and (D) a combination square. This last square consists of a graduated steel rule with an accurately machined head. The two edges of the head provide for measurements of 45° and 90°.

The usual sizes of try squares have blades ranging 3 to 15 inches long. The stock is approximately $\frac{1}{2}$-inch thick, with the blade-inserted midway between the sides of the stock. The stock is made thicker than the blade so that its face may be applied to the edge of the wood and the steel blade may be laid on the surface to be marked. Usually the blade is provided with a scale of inches divided into eighths.

Miter and Combined Try-and-Miter Squares

The term *miter*, strictly speaking, signifies any angle except a right angle, but, as applied to squares, it means an angle of 45°.

In the miter square, the blade (as in the try square) is permanently set but at an angle of 45° with the stock (see Figure 3-7).

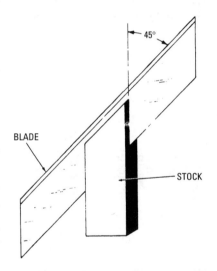

Figure 3-7 A typical miter square. It differs from the ordinary try square in that the blade is set at an angle of 45° with the stock, and the stock is attached to the blade midway between its ends.

A try square may be made into a combined try-and-miter square when the end of the stock to which the blade is fastened is faced off at 45°, as along the line *MS* in Figure 3-8. When the 45° face (*MS*) of the stock is placed against the edge of a board, the blade will be at an angle of 45° with the edge of the board (see Figure 3-9).

Figure 3-10 shows an improved form of the combined try-and-miter square. Because of the longer face (*LF*), as compared with the short face (*MS*) in Figure 3-8, the blade describes an angle of 45° with greater precision. Its worst disadvantage is that it is awkward to carry because of its irregular shape. However, its precision greatly outweighs its disadvantage.

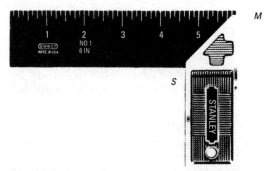

Figure 3-8 A combined try square and miter square. Because of its short 45° face (MS), it is not as accurate as the miter square, but it answers the purpose for ordinary marking and eliminates the necessity for extra tools.

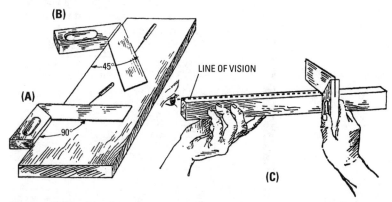

Figure 3-9 Combined try-and-miter square as used for 90° marking at A, and 45° marking at B. At C the try square is used to test squareness of edge face with side of board. The square should be placed in several positions along the edge. Should light show under the blade, it indicates that the surface is not at 90° with the side of the board or square at that section, and all such places should be trued up with a plane.

A square having a blade not exactly at the intended angle is said to be out of true, or simply out, and good work cannot be done with a square in this condition. A square should be tested and if found to be out, should be returned.

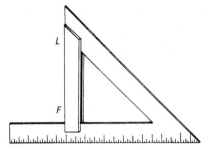

Figure 3-10 An improved form of the combined try-and-miter square.

Figure 3-11 and Figure 3-12 show the method of testing the square. This test should be made not only at the time of purchase but frequently afterward, because the tool may become imperfect from a fall or rough handling.

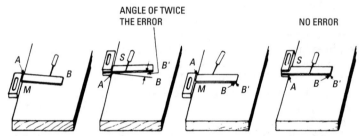

Figure 3-11 Method of testing a try square. If the square is out (angle at 90°), scribed lines AB and AB′ for positions M and S of the square (left side) will not coincide. Angle BAB′ is twice the angle of error. If the square is perfect, lines AB and AB′ for positions M and S will coincide (right side).

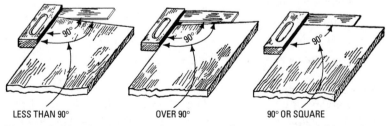

Figure 3-12 Application of try square for testing end of board to determine if the cross-cut is square with longitudinal edge of board.

Under no circumstances should initials or other markings be stamped on the brass face of the ordinary try square because the burrs that project from bending the brass face will throw the square out of truth. For this reason, manufacturers will not take back a square with any marks stamped on the brass face.

Framing or Steel Square

Although the *framing square* is commonly called the *steel square*, this is not entirely accurate because all types of squares may be obtained that are made entirely of steel. It is properly called a framing square because with its framing table and various other scales, it is adapted especially for use in house framing, although its range of usefulness makes it valuable to any woodworker (see Figure 3-13).

The framing square consists of two essential parts: the tongue and the body (or blade). The *tongue* is the shorter, narrower part, and the *body* is the longer, wider part. The point at which the tongue and the body meet on the outside edge is called the *heel*.

There are several grades of squares, including polished, nickel-plated, blued, and royal copper. The blued square with figures and scales in white is perhaps the most desirable. A size that is widely used has an 18-inch body and a 12-inch tongue. However, many uses require the largest size. The larger size has a body that measures 24×2 inches and whose tongue measures 16 (or 18) $\times$ $1\frac{1}{2}$ inches.

The feature that makes this square so valuable a tool is its numerous scales and tables:

- Rafter or framing table
- Essex table
- Brace table
- Octagon scale
- Hundredths scale
- Inch scale
- Diagonal scale

Rafter or Framing Table

This is always found on the body of the square. It is used for determining the length of common valley, hip, and jack rafters, and the angles at which they must be cut to fit at the ridge and plate. This table appears as a column six lines deep under each inch graduation from 2 to 18 inches. Figure 3-14A shows only the 12-inch section of this table. At the left of the table will be found letters indicating the

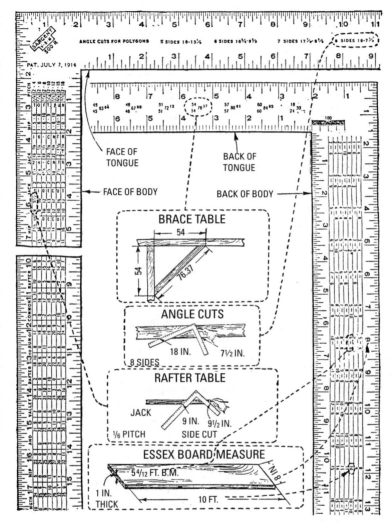

Figure 3-13 The front and back views of a typical framing square.

application of the figures given. Multiplication and angle symbols are applied to this table to prevent errors in laying out angles for cuts.

Essex Table

This is always found on the body of the square, as shown in Figure 3-14B. This table gives the board measure in feet and twelfths

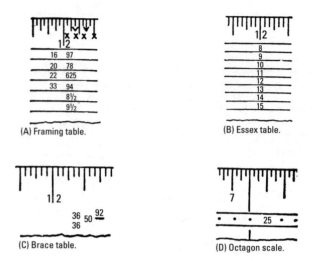

(A) Framing table.

(B) Essex table.

(C) Brace table.

(D) Octagon scale.

Figure 3-14 Typical framing-square markings.

of a foot of boards 1 inch thick (of usual lengths and widths). On certain squares, it consists of a table eight lines deep under each graduation, as seen in the figures that represent the 12-inch section of this table.

Brace Table

This table is found on the tongue of the square, a section of which is shown in Figure 3-14C. The table gives the length of the brace to be used where the rise and run are from 24 to 60 inches and are equal.

Octagon Scale

This scale is located on the tongue of the square (see Figure 3-14D) and is used for laying out a figure with eight sides on a square piece of timber. On this scale, the graduations are represented by 65 dots located $5/24$ of an inch apart.

Hundredths Scale

This scale is found on the tongue of the square. By means of a divider, decimals of an inch may be obtained. It is used particularly in reference to brace measure.

Inch Scales

On both the body and the tongue there are (along the edges) scales of inches graduated in $1/32$, $1/16$, $1/12$, $1/10$, $1/8$, and $1/4$ of an inch. Various combinations of graduations can be obtained according to the type of square. These scales are used in measuring and laying out work to precise dimensions.

Diagonal Scale

Many framing squares are provided with what is known as a *diagonal scale* (see Figure 3-15). One division (ABCD) of this scale is shown enlarged for clearness (see Figure 3-16). The object of the diagonal scale is to give minute measurements without having the graduations close together, where they would be hard to read. In construction of the scale (see Figure 3-16), the short distance AB is $1/10$ of an inch. Evidently, to divide AB into ten equal parts would bring the divisions so close together that the scale would be difficult to read. Therefore, if AB is divided into ten parts, and the diagonal BD is drawn, the intercepts 1a, 2b, 3c, and so on, drawn through 1, 2, 3, and so on, parallel to AB, will divide AB into $1/10$, $2/10$, $3/10$, and so on, of an inch. Thus, if a distance of $3/10$ AB is required, it may be picked off by placing one leg of the dividers at 3 and the other leg at c, thereby producing $3c = 3/10$ AB.

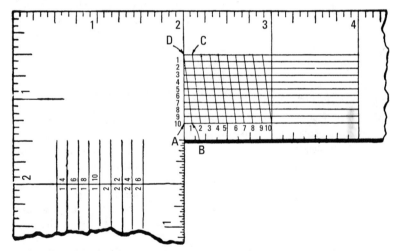

Figure 3-15 Diagonal scale on a framing square used to mark off hundredths of an inch with dividers.

Because of the importance of the framing square and the many problems to be solved with it, the applications of the square are provided in Chapter 17.

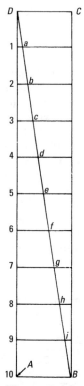

Figure 3-16 Section ABCD of Figure 3-15, enlarged to illustrate the principle of the diagonal scale.

Combination Square

This tool (see Figure 3-17), as its name indicates, can be used for the same purposes as an ordinary try square, but it differs from the try square in that the head can be made to slide along the blade and clamp at any desired place. Combined with the square, it is also a level and a miter. The sliding of the head is accomplished by means of a central groove in that a guide travels in the head of the square. This permits the scale to be pulled out and used simply as a rule. It is frequently desired to vary the length of the try-square blade. This is readily accomplished with the combination square. It is also convenient to square a piece of wood with a surface and at the same time tell whether one or the other is level, or *plumb*. The spirit level in the head of the square permits this to be done without the use of a separate level. The head of the square may also be used as a simple level.

Because the scale may be moved in the head, the combination square makes a good marking gauge by setting the scale at the proper position and clamping it there. The entire combination square may then be slid along as with an ordinary gage. As a further convenience, a scriber is held frictionally in the head by a small brass bushing. The scriber head projects from the bottom of the square stock in a convenient place to be withdrawn quickly.

In laying out, the combination square may be used to scribe lines at miter angles as well as at right angles, since one edge of the square head is at 45°. Where micrometer accuracy is not essential, the blade of the combination square may be set at any desired position, and the square may then be used as a depth gauge to measure in mortises, or the end of the scale may be set flush with the edge of the square and used as a height gauge.

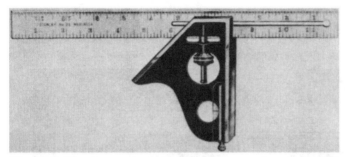

Figure 3-17 A typical combination square with a grooved blade, level, and centering attachments.

The head may be unclamped and entirely removed from the scale, and a center head can then be substituted so that the same tool can quickly be used to find the centers of shafting and other cylindrical pieces. In the best construction, the blade is hardened to prevent the corner from wearing round and destroying the graduations, thus keeping the scale accurate at all times. This combination square combines a rule, square, miter, depth gage, height gage, level, and center head. It permits work that is more rapid. It saves littering the bench with a number of tools, each of which is necessary but may be used only rarely. It tends toward the goal for which all carpenters are striving—greater efficiency. Figure 3-18 shows some of the uses for the combination square.

Sliding T Bevel

A *bevel* is virtually a try square with a sliding adjustable blade that can be set at any angle to the stock. In construction, the stock may be of wood or steel. When the stock is made of wood, it normally has brass mountings at each end, and it is sometimes concave along its length. The blade is of steel with parallel sides, and its end is at an angle of 45° with the sides (see Figure 3-19). The blade is slotted, thereby allowing linear adjustment and the insertion of a pivot (or screw pin) that is located at the end of the stock. After the blade has been adjusted to any particular angle, it is secured in position by tightening the screw lever on the pivot. This action compresses the sides of the slotted stock together, thus firmly gripping the blade (see Figure 3-20).

When selecting a bevel, care should be taken to see that the edges are parallel and that the pivot screw, when tightened, holds the blade firmly without bending it.

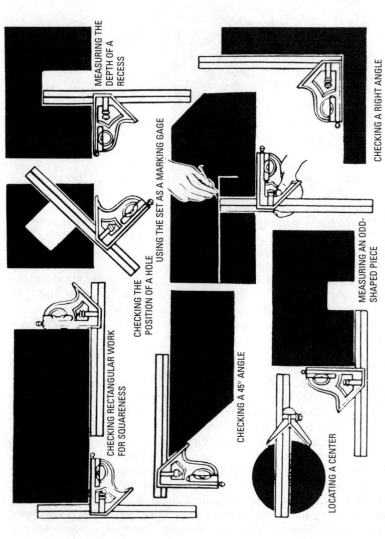

MEASURING THE DEPTH OF A RECESS

CHECKING A RIGHT ANGLE

USING THE SET AS A MARKING GAGE

CHECKING THE POSITION OF A HOLE

CHECKING RECTANGULAR WORK FOR SQUARENESS

MEASURING AN ODD-SHAPED PIECE

CHECKING A 45° ANGLE

LOCATING A CENTER

Figure 3-18 Some of the many uses of the combination square.

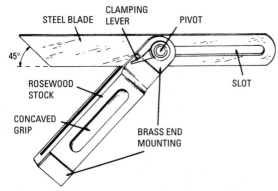

Figure 3-19 A sliding T bevel with a steel blade, rosewood stock, and brass end mountings. Since the size of a bevel may be expressed by the length of either its stock or its blade, care should be taken to specify that the dimension is given when ordering to avoid mistakes.

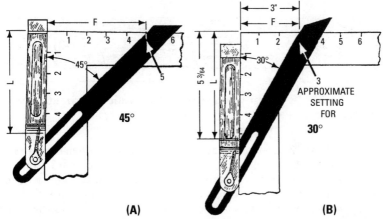

(A)   **(B)**

Figure 3-20 Method of setting the bevel to various angles by using the framing square. At A, place the stock of the bevel against one side of the square as shown and adjust both stock and blade to square until the two legs L and F of the right angle triangle thus formed are of equal length. As shown, both measure 5 inches. At B, to obtain the 30° setting, taking a value of say 3 in for F, then the exact value of L is 5.196 in or approximately $5^{13}/_{64}$. With a table of natural sines and cosines (F = sine and L = cosine), a setting for any angle may be obtained. Rule: Divide value of cosine (obtained from table) by value of sine of required angle. Multiply ratio thus obtained by any assigned value of leg F, and product will be corresponding length of leg L.

Center Square
Another useful square that is of recent vintage is the *center square*. This works like a protractor and can be used to find the center of any size circle quickly and easily, as well as to determine right angles quickly.

Level
The level is used for both guiding and testing—to guide in bringing the work to a horizontal or vertical position and to test the accuracy of completed construction. It consists of a long rectangular body of wood or metal that is cut away on its side and near the end to receive glass tubes that are almost entirely filled with a nonfreezing liquid with a small bubble free to move as the level is moved (see Figure 3-21).

LEVEL GLASS   PLUMB GLASS

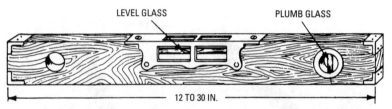

|← 12 TO 30 IN. →|

Figure 3-21   A typical wooden spirit level with a horizontal and a vertical tube.

The side and end tubes are at right angles so that when the bubble of the side tube is in the center of the tube, the level is horizontal. When the bubble of the end tube is in the center, the level is vertical. Hence, when the level is placed on a surface, its levelness (or plumbness) can be tested. Levels also come with magnetic edges that can make placement easier (see Figure 3-22).

Plumb Bob
The word *plumb* means perpendicular to the plane of the horizon, and since the plane of the horizon is perpendicular to the direction of gravity at any given point, gravity is used to determine vertical lines by the device known as a plumb bob.

This tool consists of a pointed weight attached to a string. When the weight is suspended by the string and allowed to come to rest (see Figure 3-23), the string will be plumb (vertical). The ordinary top-shaped solid plumb bob is problematic because of a

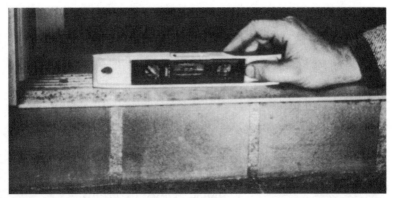

Figure 3-22 Levels come in a variety of sizes. This small torpedo type is handy for judging the level when the area is not large enough to accommodate a regular-length tool. *(Courtesy of Stanley)*

too-blunt point and not enough weight. For outside work, the matter of weight is important, since when the plumb bob is used with a strong wind blowing, the excess surface presented to the wind will magnify the error. To reduce the surface for a given weight,

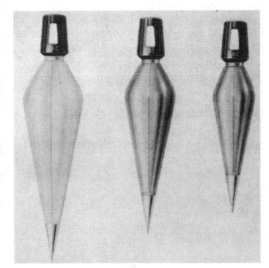

Figure 3-23 A solid plumb bob.

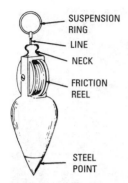

SUSPENSION
RING

LINE

NECK

FRICTION
REEL

STEEL
POINT

Figure 3-24   An
adjustable plumb bob.

the bob is bored and filled with mer-
cury. Figure 3-24 shows an adjustable
bob with a self-contained reel on which
the string is wound.

Summary

Guiding and testing tools are invaluable to the carpenter because without
them it would be impossible to mark
material with the precision that is often demanded. A good variety of these
tools should be in every toolbox and
should be used frequently.

One of the more important guiding
and testing tools is the straightedge,
which can be of metal or wood and can be used to guide the pencil
or scriber. It also comes in a variety of lengths.

The try square is used to check for right-angle cuts on any
straight-edged material. There are various types of try squares, such
as double try squares, combination try squares, and try-and-miter
squares.

A framing square (sometimes called a steel square) is a square
with framing tables and various other scales. It is adapted especially
for use in house framing. The features that make this square such
a valuable tool are the table of rafter measurements for common,
valley, hip, and jack rafters, and the angles at which they must be
cut to fit the ridge and plate.

The miter box is a tool used to guide the saw in cutting material,
generally at 45° and 90° angles. A miter box has at least two 45°
angles for cutting right and left miters, but many miter boxes allow
other angle cuts.

The level is a tool used to bring the work to a true horizontal or
vertical position. By holding the level on a surface that is horizontal
or vertical, the tool itself may be checked for accuracy.

Review Questions

1. What is a straightedge?
2. Explain the use of the plumb bob and the level.
3. Name a few of the common types of squares.
4. What is a shooting board?

5. What does the word *plumb* mean? What is a plumb bob used for?
6. What is the other most commonly used name for the framing square?
7. How many 45° angles does the miter box contain?
8. What kind of tool is a sliding T bevel? What does it do?
9. What is the Essex table?
10. How does the try square differ from the steel square?

Chapter 4

Layout Tools

In good carpentry and joinery, a great deal depends on accuracy in laying out the work. The term *laying out* means the operation of marking the work with a tool (such as a pencil or scriber) so that the various centers and working lines will be set off in their proper relationship. These lines are followed by the carpenter in cutting and other tooling operations necessary to bring the work to its final form.

In laying out, the guiding tools just described are used to guide the pencil or scriber. The measurements are made by the aid of the measuring devices described in a later chapter.

According to the degree of precision required in laying out, the proper marker to use is as follows:

- *For extremely rough work*—The chalk box and reel or the carpenter's pencil with rectangular lead.
- *For rough work*—The lead pencil with round lead.
- *For accurate work*—The scratch awl.
- *For precision work*—The scriber or knife.

For efficiency, a good degree of common sense should be used in deciding which marker to use. Thus, it would be self-defeating to use a machine-hardened steel scriber with a needlepoint to mark off rafters, or to use a carpenter's pencil with an acre of soft lead on the point to lay out a fine dovetail joint.

Chalk Box and Line

The chalk line (see Figure 4-1) is used to mark a long straight line between two points that are too far apart to permit the use of a square or straightedge.

The chalk box is usually constructed of aluminum or plastic. Inside the box is a reel that is fitted with a lightweight string or cord; the box has powdered chalk in it. To use it, the line is stretched between two points. When the string is taut, it is pulled up and released, thus leaving a chalk line on the surface. Note the right way and the wrong way to use the chalk line shown in Figure 4-2.

Carpenter's Pencil

The conventional carpenter's pencil (which is rectangular in cross-section) is considerably larger than an ordinary pencil. The idea in

Figure 4-1   A chalk box with 100 feet of string.

making the lead this shape is to permit its use on rough lumber without too frequent sharpening and to give a well-defined, plainly visible line. Because of the thickness of the line, the carpenter's pencil is not intended for fine work, but is used principally for marking boards and so on that are to be sawed.

Figure 4-3 shows a section of the carpenter's pencil. When marking with a carpenter's pencil, the mark must be made in the direction of the long axis of the lead as shown in Figure 4-4B, and not as shown in Figure 4-4C. The proper method of sharpening the pencil, as shown in Figure 4-4A, should be noted.

Ordinary Pencil

Ordinary pencils with thin leads are, of course, also used in carpentry. Since the lead is smaller than that of the carpenter's pencil, it produces a finer marking line and is used on smooth surfaces where more accurate marking is required than can be obtained with the carpenter's pencil. When using, the best results are obtained by twisting the pencil while drawing the lines to retain the conical shape the lead gets when sharpened.

Marking or Scratch Awl

This tool consists of a short piece of round steel that is pointed at one end with the other end permanently fixed in a convenient handle (see Figure 4-5). A scratch awl is used in laying out fine work where a lead pencil mark would be too coarse for the required degree of precision.

Scriber

A scriber is a tool of extreme precision and, although intended especially for machinists, it should be in the tool kit of carpenters who are engaged in very fine work.

A scriber is a hardened steel tool with a sharp point designed to mark extremely fine lines. The most convenient form of scriber is the pocket (or telescoping) type (see Figure 4-6). Its construction renders it safe to carry in the pocket.

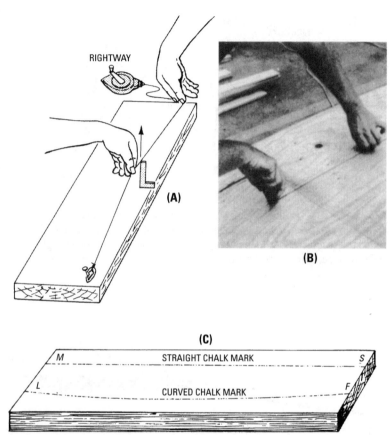

Figure 4-2 The right way to use the chalk box. When pulling up the line, always do so in a direction that is at right angles with the board. If the chalk line is pulled straight up, as in A, a straight chalk mark MS will be obtained; if the line is pulled up to one side, a curved line LF will be produced.

Figure 4-3 Section of a typical carpenter's pencil.

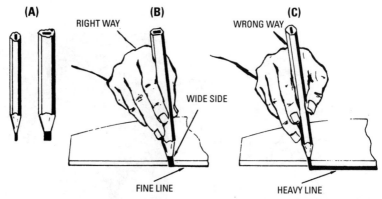

Figure 4-4 The right and wrong ways to use the carpenter's pencil. (A) Side and end views of a carpenter's pencil. (B) A fine line is obtained with the long side of the lead turned in the direction of the straight edge. (C) A wide, undefined line is produced when the pencil is used in this position.

Figure 4-5 An ordinary scratch awl with a forged blade and a hardwood handle.

Figure 4-6 A telescoping scriber in the open and closed positions.

Compass and Dividers

The compass is an instrument used for describing circles or arcs by scribing. It consists of two pointed legs that are hinged firmly by a rivet. This is to remain set in any position by the friction of the hinged joint. The usual form of carpenter's compass is shown in Figure 4-7. It should not be used in place of dividers for dividing an arc or line into a number of equal divisions because it is not a tool of precision.

Figure 4-7 A typical compass.

The difference between dividers and compasses is that the dividers are provided with a quadrant wing projecting from one of the two hinged legs through a slot in the other leg. A setscrew on the slotted leg enables the instrument to be securely locked to the approximate dimension. Then it is adjusted with precision to the exact dimension by a screw at the other end of the wing. A spring pressing against the wing holds the leg firmly against the screw (see Figure 4-8). Because of the wing, the tool is frequently called *winged dividers*.

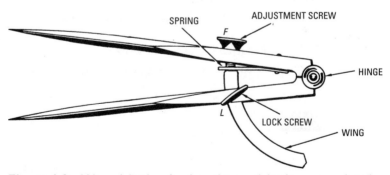

Figure 4-8 Winged dividers for describing and dividing arcs and circles. When the dividers are locked in the approximate setting by lock screw L, the tool can be set with precision to the exact dimension by turning adjustment screw F, against which the leg is firmly held by the spring to prevent any lost motion.

Summary

Accuracy in carpentry work depends on the correct use of good tools. In layout work, the guiding tools are used to guide a pencil or scriber.

The conventional carpenter's pencil has a rectangular lead that is considerably larger than the lead in an ordinary pencil. With this design, the pencil can be used frequently without sharpening the lead.

The scratch awl is a short piece of round steel that is pointed at one end with the other end permanently fixed in a convenient handle. A scratch awl is used in laying out fine work where a pencil would be too coarse for the required precision.

The compass or divider is an instrument used for describing circles or arcs. It is designed with two pointed legs that are hinged at one end.

Review Questions

1. What is the difference between a divider and a compass?
2. Why is a carpenter pencil lead rectangular?
3. What is the purpose of the scratch awl?
4. What marking tool would be used for precision work?
5. What is a chalk line and how is it used?
6. How does good common sense become a factor in laying out where to cut wood?
7. For what is a chalk box used?
8. How does a carpenter's pencil differ from an ordinary writing pencil?
9. How is the scriber used in carpentry work?
10. What is the difference between a scratch awl and a nail set?

Chapter 5

Rules, Scales, and Gages

In laying out work, after having scribed a line with one of the marking tools described in Chapter 4 and aided by a guiding tool, the next step is usually to measure off on the scribed line some given distance. This is done with a suitable *measuring tool.*

Rules

The most basic tool for the carpenter is the *ruler*, or *rule*. Most carpenters favor the 16-foot-long steel tape (see Figure 5-1). It enables them to take both short and long measurements with accuracy.

Figure 5-1   Most carpenters favor 16-foot steel tape. It can be used for long or short measuring jobs. *(Courtesy of The American Plywood Assn.)*

Folding Wood Rule

Another rule (an old favorite, and deservedly so) is the *folding wooden rule.* This commonly comes 6 feet long and in 1-foot increments. The rule is hinged together and folds out for use (see Figure 5-2). Such a ruler is invaluable because it permits one-hand

Figure 5-2 The folding rule is another old favorite. One advantage of the rule is that it can be used with one hand. *(Courtesy of Vaughn & Bushnell)*

measuring of something. It has the rigidity required for laying it across large areas (see Figure 5-3).

Figure 5-3 The folding rule is also good for stretching across an area. It has the required rigidity.

A variation on the basic wood rule is the wood rule with an extension that pulls out of one end. This is useful when you are taking an inside measurement (see Figure 5-4). Just slide the extension out to measure the last few inches.

Other Rules

The *bench rule* is another handy rule. It is mounted on the edge of a bench and things are held against it for measuring.

A 6-inch *caliper rule* is also good. It has a hooklike part that slides in a groove along the rule. The outside caliper rule is excellent for checking the thickness of round stock (see Figure 5-5). You can also get caliper rules for checking the widths of slots and grooves. Such rules are known as *inside caliper rules*.

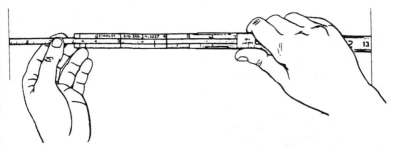

Figure 5-4 The folding rule with extension is good when regular segments are too big for measurement.

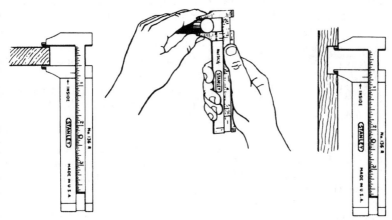

Figure 5-5 An outside caliper (left and center) can be used to measure the outside of square or round objects. An inside caliper (right) is for measuring widths of grooves and the like.

The big idea when buying a rule (as with other types of tools) is to buy quality. A saw that cuts in an errant way may be tolerated for the duration of a job, but a measuring tool that isn't accurate can turn a job into a disaster.

Using Rules
The axiom when using any kind of measuring device is to measure twice, cut once. Another notion, not so well known, is to use the same brands of rules if you are working with someone. What constitutes 12 feet to one manufacturer may not mean the same thing to another, even though both produce quality tools. If two brands

of rules, then, differ by only $1/32$ inch per foot, this may not mean much in a short measuring job, but it can spell big problems as the footage goes up.

When measuring anything, hold the rule on its edge so that the measuring marks are in contact with the work (see Figure 5-6). Make the mark with a sharp knife or pencil.

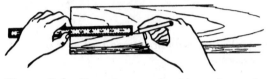

Figure 5-6   To make a mark with a rule on the end of a board, place the rule up so that the marks are on the board, and then mark.

Figure 5-7 To measure the end of a board, use your forefinger to guide one end of the rule flush, then use your thumb as shown to fix the mark.

If you are measuring the end of a board, use a forefinger to guide one end of the rule into alignment with one side of the board and a thumbnail to fix the exact measuring mark on the other side (see Figure 5-7).

If you want to make a line parallel to the edge of a board, hold the rule as shown in Figure 5-8, guiding the rule along with the pencil in the position shown.

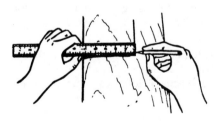

Figure 5-8   To make a mark parallel to the edge, hold the rule as shown, place the pencil on the end of the rule, and pull along.

Lumber Scale

When estimating lumber, a lot of time is saved in the lumberyard by using a *lumber scale* from which the board feet measure may be read off directly. This scale gives an approximate result. When using the scale, it is customary to read to the nearest figure and, when there is no difference, to alternate between the lower and then the higher figure on different boards. Figure 5-9 shows a board scale graduated for boards of 12-, 14-, and 16-foot lengths and the method of using the scale.

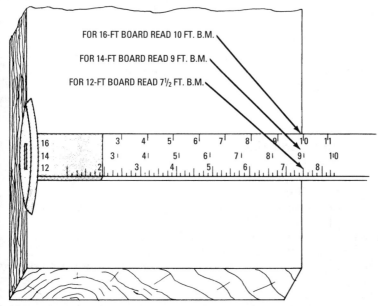

Figure 5-9 A typical board rule that is used to measure lumber in board-measure units. To use the rule, place the head of the rule against one edge of the board, and read the figure nearest the other edge (width) of the board in the same line of figures on which the length is found. This reading will give the number of feet board measure for the piece of lumber being measured.

There are many types of lumber scales in use in various sections of the United States and Canada.

Spring Steel Board Rule

This rule is made of tempered spring steel so that it will bend to the board and, when released, will return exactly straight. It is provided with a wood handle and a leather slide for handling the rule at any part of the blade. Types of rules include the following:

- 3-tier
- $3^1/_2$-foot inspector's rule

- 3-tier, 3-foot board rule
- 3-tier, 2½-foot sorting rule

All three types are marked on one side to measure 8, 10, and 18 feet. The opposite side is marked 12, 14, and 16 or 18, 20, and 22 feet.

Marking Gages

Tools of this type are used to mark a piece of wood that is to be sawed or otherwise tooled. Following are some of the several types of marking gages:

- Single bar
- Double bar
- Single bar with slide
- Butt

Single-Bar Gage

This is used for making a single mark (such as for sawing). It consists of a bar with a scriber (or pin) at one end with a scale that is graduated in inches and sixteenths. The bar passes through a movable head that may be clamped at any distance from the scriber point (see Figure 5-10). Figure 5-11 shows how to use a marking gage.

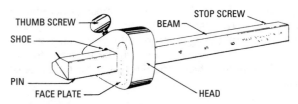

Figure 5-10 A single-bar marking gage. It is fitted with a head, faceplate, and thumbscrew and is provided with a scale that is graduated in inches and sixteenths.

Double-Bar Gage

This type of gage is designed especially for mortise marking. There are two independent bars working in the same head. One pin is affixed to each bar. After setting the bars for the proper marking of the mortise, one side is marked with one bar, and the gage is then turned over for marking the other side (see Figure 5-12).

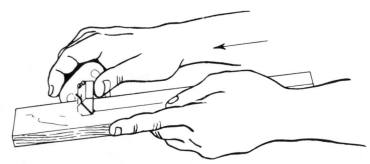

Figure 5-11 The method of using a marking gage. When setting the gage, use a rule, unless it is certain that the scriber point is located accurately with the graduations on the bar. When marking, the gage should be held as indicated. The face of the head is pressed against the edge of the board. Care must be taken to keep it true with the edge so that the bar will be at right angles with the edge and the line scriber will be at the correct distance from the edge. The line is usually scribed by pushing the gage away from the worker. Always work from the face side, as shown.

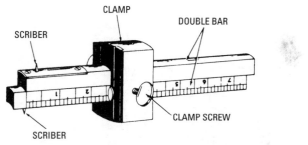

Figure 5-12 A double-bar marking gage. It is used for marking a given distance between two parallel lines and a given distance from the edge of a board. As shown in the illustration, each bar has a setscrew for clamping it in any desired position.

Slide Gage

One objection to the double-bar gage is that two operations are required that can both be performed with a slide gage in one operation. The underside of the bar is provided with a flush slide having a scriber B at the end of the slide, with another scriber A at the end of the bar (see Figure 5-13). These two scribers, when set to the required distances from the head, mark both sides of the tenon

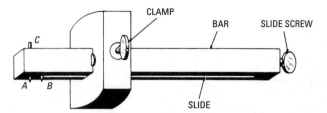

Figure 5-13 A slide marking gage. The bar has a scriber (C) on the upper side for single marking and a scriber (A) on the lower side that, with the scriber (B) on the slide, works flush in the bar. The distance between scriber points A and B is regulated by the slide screw at the end of the bar.

or mortise to the size required with just one stroke (see Figure 5-14). On the upper side, there is one scriber C for single marking.

Butt Gage

When hanging doors, three measurements must be marked:

- The location of the butt on the casing

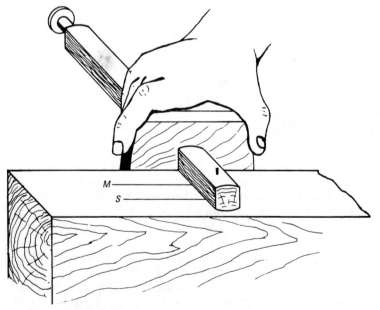

Figure 5-14 The method of using a slide gage when marking a mortise. Note that the marks (M and S) for the sides of the mortise can both be scribed in one operation.

- The location of the butt on the door
- The thickness of the butt on the casing

A butt gage is a type of gage having three cutters that are purposely arranged so that no change of setting is necessary when hanging several doors. In reality, these tools comprise rabbet-gages, marking-gages, and mortise-gages of a scope sufficient for all door trim (including lock plates, strike plates, and so forth).

Figure 5-15 shows a typical butt gage. The cutters are mounted on the same bar and are set by one adjustment with the proper allowance for clearance. When casings have a nailed-on strike instead of being rabbeted, a marking gage that will work on a ledge as narrow as $1/8$ inch is required. In this case, the same distance is marked from the edge of the casing and from the edge of the door that is not engaged when closing. Certain gages can be used on such work. One cutter marks the butt, and one cutter marks its thickness. Other gages are made so that they can be used as inside or outside squares for squaring the edge of the butt on either the door or the jamb.

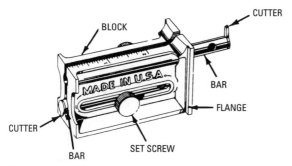

Figure 5-15 A typical butt gage. It is used to mark the location of the butt on casings and doors. Three separate cutters, one for each dimension, eliminate changing the setting when more than one door is hung. It is also used as a marking and mortising gage and as an inside and outside square for squaring the edge of a butt on the door and jamb. It is graduated in sixteenths for 2 inches.

Summary

The most common measuring tool known in any type of work is the rule. There are many different types of rules, but in this type of work, it is referred to as the carpenter's rule. One familiar type is the 6-foot folded wooden rule. The most popular rule is the 16-foot steel tape.

Marking gages are used to mark pieces of wood that are to be sawed. There are several types of marking gages.

Review Questions

1. What type of rule is most popular?
2. What is a marking gage? Explain how it works.
3. How do you use a rule to mark a line parallel to a board?
4. The folding rule is also good for stretching_____an area.
5. Where is the bench rule mounted?
6. True or false: The old axiom says, measure twice, cut once.
7. A typical_____rule is used to measure lumber in board-measure units.
8. List four types of marking gages.
9. What is a butt gage? Where is it used primarily?
10. How is a sliding gage used to mark a mortise?
11. The most common measuring tool known in any type of work is the_____.

Chapter 6

Clamps, Vises, and Workbenches

An essential part of the shop equipment necessary for good carpentry is a proper assortment of *holding tools*. Some tooling operations require that the work be held rigid, even when considerable force is applied, such as in planing and chiseling.

The workbench, considered broadly with its attachments, may be called the main holding tool, and unless this important part of the equipment is constructed amply substantial and rigid, it will be difficult to do good work.

Holding tools may be generally classed as supporting tools and retaining tools. When marking or sawing, it is usually only necessary to support the work by placing it on the bench or on sawhorses. However, in planing, chiseling, and some nailing operations, the work must not only be supported but also held rigidly in position.

Horses or Trestles

Horses, or *trestles*, are used in various ways to simply support the work when it is of such large dimensions that the bench cannot conveniently be used (especially for marking and sawing planks). No shop equipment list is complete without a pair of sawhorses. A sawhorse, as usually made, consists of a 3- or 4-foot length of 2-inch × 4-inch or 2-inch × 6-inch stock for the cross beam, with a pair of 1-inch × 3-inch or $1^{1}/_{2}$-inch × 4-inch legs at each end, depending on the expected weight of the work.

The height of the sawhorse is usually 2 feet (see Figure 6-1). Note that the legs are inclined outward both lengthwise and crosswise, and a problem arises as to how to determine the length of the legs having this double inclination for a given height of sawhorse (see Figure 6-2 and Figure 6-3).

Clamps

Frequently it is necessary to press pieces of wood together tightly that are mortise and tenon, tongue-and-groove, or simply glued. The bench vise is not always convenient, or is may be required for other work. Then *clamps* are used.

C-Clamp

These come with jaws that open to 12 inches wide. As the name suggests, they are in the shape of the letter C, and are very useful for clamping a wide variety of items. Some C-clamps come with shielded screws to protect against hammer blows (see Figure 6-4).

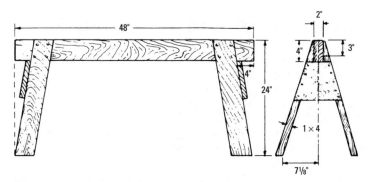

Figure 6-1 Side and end views of a typical carpenter's sawhorse with dimensions suitable for general use.

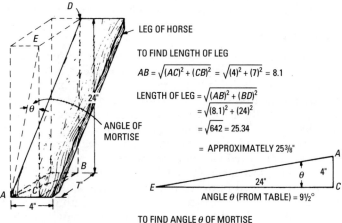

LEG OF HORSE

TO FIND LENGTH OF LEG

$$AB = \sqrt{(AC)^2 + (CB)^2} = \sqrt{(4)^2 + (7)^2} = 8.1$$

$$\text{LENGTH OF LEG} = \sqrt{(AB)^2 + (BD)^2}$$
$$= \sqrt{(8.1)^2 + (24)^2}$$
$$= \sqrt{642} = 25.34$$
$$= \text{APPROXIMATELY } 25\frac{3}{8}"$$

ANGLE θ (FROM TABLE) = 9½°

TO FIND ANGLE θ OF MORTISE

$$\text{TAN } \theta = \frac{AC}{CE} = \frac{4}{24} = 0.166$$

Figure 6-2 A method of finding the length of a sawhorse leg and the angle, or inclination of side, of the mortise for the leg. To find Φ by calculation, a table of natural trigonometric functions is necessary.

Deep-Throat Type

Closely aligned to the regular C-clamp is the deep-throat type. It is good when you need to slip the clamp over the edge of something to clamp it on the interior where other clamps would have difficulty reaching.

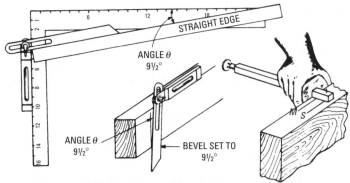

ANGLE θ
9½°

STRAIGHT EDGE

ANGLE θ
9½°

BEVEL SET TO
9½°

M S

Figure 6-3 The method of setting the bevel angle (Φ) with the aid of a square and a straightedge. Place the straightedge on the square so that one side of the right triangle thus formed will be 24 inches (height of the sawhorse) and the other side will be 4 inches (distance to edge of leg from end of beam). Place the blade of the bevel against the straightedge, and place the stock against the side of the square. Clamp the bevel to this angle, which is the proper slope for the side of the mortise.

Figure 6-4 C-clamps hold a miter joint. The blocks of dark wood are temporarily glued on before clamping. After glue in the miter joint is set, the clamps are unscrewed and the blocks are planed or sanded off.

Edge Clamp

There are a few kinds of edge clamps, one of which is basically a C-clamp with another screw set in at a right angle to the jaw opening (see Figure 6-5).

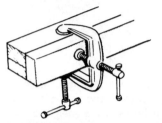

Figure 6-5 Edge, or corner, clamp, good for securing stock to edges.

(Courtesy of Hand Tool Institute)

Pipe Clamp

Separate fittings that comprise the clamp are bought separately and are fastened to $1/4$- or $3/4$-inch iron pipe of whatever length is desired. This type of clamp can be used to clamp very deep work.

Steel-Bar Clamp

This resembles the bar clamp (see Figure 6-6). It is available with openings from 2 feet to 8 feet.

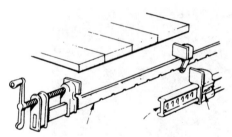

Figure 6-6 Bar clamp, good for wide work.

(Courtesy of Hand Tool Institute)

Spring Clamp

These are like having an extra hand. They are like oversized clothespins and can be used to apply light pressure (see Figure 6-7).

Hand Screw

Hand screws are useful for face-to-face gluing. They keep the pressure evenly applied to the surfaces (faces) coated with glue. There is a right way and a wrong way to adjust the clamps to ensure even pressure on the gluing surface (see Figure 6-8).

Band Clamp

This is for clamping large, irregularly spaced items (see Figure 6-9). There is a strip of nylon or canvas webbing 1 to 2 inches wide and 15 feet long. The ratchet head is tightened securely by wrench or handle.

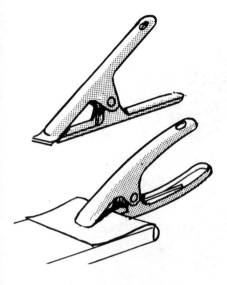

Figure 6-7 Spring clamp, good for light duty.

(Courtesy of Hand Tool Institute)

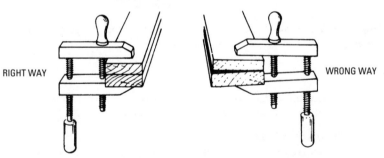

RIGHT WAY

WRONG WAY

Figure 6-8 Right and wrong ways to use a hand screw. First, set the jaws to approximately the size of the material to be clamped. When placing the hand screw on the work, keep the points of the jaws slightly more open than the outer ends. Final adjustment of the inside screw will then bring the jaws exactly parallel, which is the proper position for clamping parallel work. Of course, if the work itself is wedge-shaped, the clamp jaws should conform so that equal pressure is applied at all points of contact. Since the screws are made of wood instead of iron or steel, proper adjustment must be used with respect to the applied pressure.

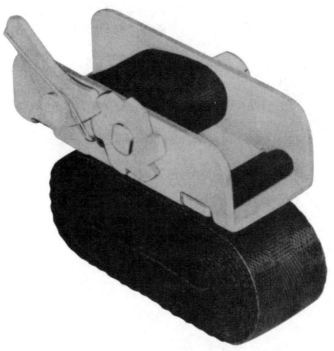

Figure 6-9 A band clamp that is used to clamp irregularly shaped objects.

Vises

Vises are permanently attached to a workbench or table in most cases. There are a number of useful styles of vises.

Woodworker's Vise

The woodworker's vise (see Figure 6-10) has two parallel flat jaws varying in size from 3 inches × 6 inches to 4 inches × 10 inches. Jaws open from 6 inches to up to 12 inches on professional models. The best position for a woodworker's vise is to have the tops of the jaws flush with the workbench top.

One feature that signifies a better vise is a fast-acting screw that allows the front jaw of the vise to be quickly set in a gripping position prior to a tightening action, something accomplished by rotating a sliding pin handle that turns a screw that applies the force. The jaws are never barefaced; wood or composition inserts are installed to protect the work-piece.

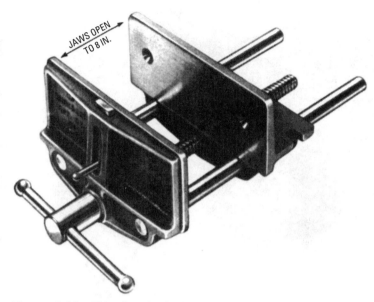

Figure 6-10 Woodworker's vise. *(Courtesy of Brink & Cotton)*

The woodworker's vise can be used to hold a wide variety of things, from boards to sheets of plywood.

One other good feature of the woodworker's vise is a spring-loaded dog in the front jaw. This can be raised to butt against one side of a very wide workpiece so that the piece is much wider than the normal jaw opening can be held.

Bench Vise

The bench vise has either a fixed base that is bolted to the bench top or a swivel base that allows the vise to be turned 220° to 360°. No matter which angle is used (220° or 360°), the main screw can be turned to open the vise as required. Jaws may be opened from 3 to 6 inches depending on the vise. Construction varies. Vises are made of gray iron, malleable iron, or steel. Machinist vises, which get rugged use, are usually made of steel or malleable iron and are more expensive than bench vises (see Figure 6-11).

Good bench vises usually include a set of pipe jaws for working pipe. Better vises also have hardened steel and serrated jaw inserts that can be used for holding pipe.

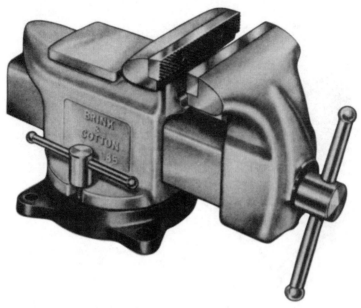

Figure 6-11   Machinist's vise. *(Courtesy of Brink & Cotton)*

Vises should be mounted so that the jaws extend over the edge of the bench. This way, long objects can be held vertically without interference.

Clamp-On and Sawhorse Vises
Two other useful vises are the clamp-on and the sawhorse. *Clamp-on vises* are bench vises, but with jaws that open only to 3 inches. As the name suggests, they clamp onto the bench. They are commonly used for light work.

The sawhorse vise is a variation of the clamp-on bench vise but is designed to clamp to a sawhorse. The sawhorse vise is L-shaped and is good for holding a board or piece of plywood in a vertical position as work is done on it.

Workbenches
To properly perform many of the numerous operations in carpentry, a suitable workbench is essential. A bench is really just a specialized tool for holding materials conveniently during bench work. However, it's far more than a simple table. The variety of clamping and

tool storage features built into an effective bench turn it into an important piece of equipment.

A rickety bench that jiggles with each stroke of work (or slowly creeps across the floor) cuts down on the efficiency of handwork. Each movement of the bench (even if very slight) means that part of every chisel or plane stroke is wasted because it is absorbed by the bench instead of transferred to the work. A solid, steady bench improves efficiency by eliminating wasted effort. This improves accuracy and the quality of work.

The height of the bench depends on the work to be done and the person doing it. A higher work surface of 36 to 42 inches is appropriate for light work. A lower surface of 30 to 36 inches is appropriate for heavy work. Consider the height of the person using the bench. A bench that is too low or too high can cause serious back pain and injury during long or repeated use. In general, carpenters' benches are made 33 inches high, whereas those for cabinetmakers are 2 to 4 inches higher.

Bench Types

Types of workbenches include the following:

- Temporary site bench
- Shop bench
- Portable bench

Temporary Site Bench

You can easily construct a bench right at the construction site (see Figure 6-12). This makes sense on large jobs, where the cost of time and materials to build the bench will be returned because of improved efficiency compared to working without a bench. Fasten the parts together with wallboard screws and a portable drill. Then the parts can be disassembled with little damage and used for materials at the end of the job or stored until needed and reassembled on the next site.

The main advantage of a temporary bench is that it is built quickly and at little expense.

Shop Bench

A permanent bench with specialized features is more suitable for the workshop (see Figure 6-13). The thick, massive bench top resists the forces applied in handwork. Often, shop benches are fastened to the floor and wall to make them absolutely stable and rigid.

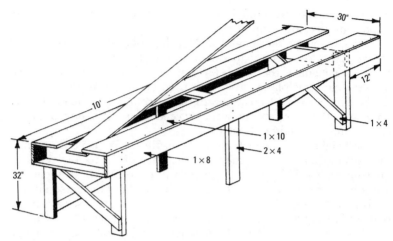

Figure 6-12 A quickly constructed workbench for temporary use. Lumber required: five 2 inch × 4 inch × 8 feet; one 1 inch × 4 inch × 16 feet; two 1 inch × 8 inch × 10 feet; and three 1 inch × 10 inch × 10 feet. The center legs add rigidity, but on a short bench they are not necessary.

Portable Bench

Manufacturers offer several styles of innovative workbenches (see Figure 6-14). These devices are lightweight and fold quickly for easy transport and storage. Although the work surface is limited in size, it has a wide variety of clamping features. The light weight provides no massive resistance to the forces of handwork, but a step or flat base is provided to hold the bench stable during heavy work. This is awkward for some operations, such as hand planing a long board. These benches are best suited for detail and finish work.

Only the largest, heavy-duty models will hold up to the rigors of professional carpentry. Even these are best suited for only light to medium duty work and may wear out after several years.

Features

Following are some of the features of common workbenches:

- Vises
- Support pegs
- Tool drop

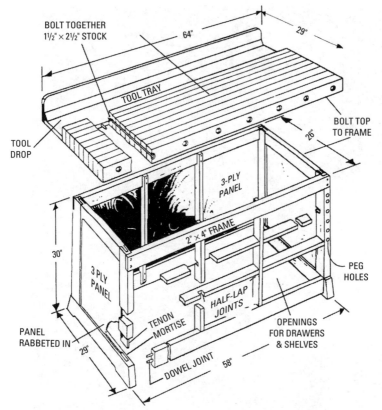

BOLT TOGETHER
1½" × 2½" STOCK

64"

29"

TOOL TRAY

TOOL DROP

BOLT TOP TO FRAME

26"

3-PLY PANEL

2" × 4" FRAME

30"

3 PLY PANEL

PEG HOLES

TENON MORTISE

HALF-LAP JOINTS

OPENINGS FOR DRAWERS & SHELVES

PANEL RABBETED IN

29"

DOWEL JOINT

58"

Figure 6-13 Construction details of a workbench designed for appearance as well as convenience. The 2-inch × 4-inch rails, corner posts, and base members provide a sturdy bench.

- Storage
- Bench stop
- Bench hook

Vises

Usually there is a large main vise at the left end of the bench at the front edge (see Figure 6-15) and a smaller vise on the right end for detail work. Face iron vises with wood or leather coverings to prevent marking or denting the lumber, especially when soft woods are used.

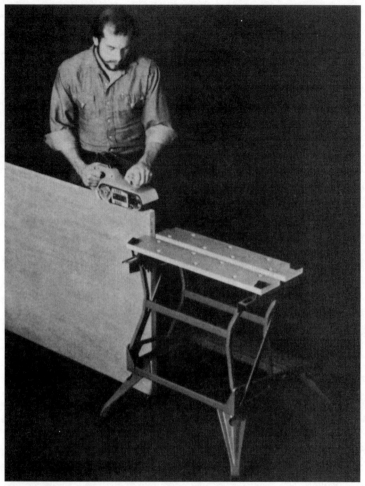

Figure 6-14 The workmate portable bench. *(Courtesy of Black & Decker)*

Support Pegs

The function of the main bench vise is to prevent the wood from moving while being worked (such as with a plane or chisel). In these operations, pressure tends to rotate the wood in the plane of the vise jaws, which act like a pivot.

In the case of a long board, this turning force (or torque) is very large when a downward pressure is applied at the far end of the board. So the vise needs to be overtightened to compensate. To avoid

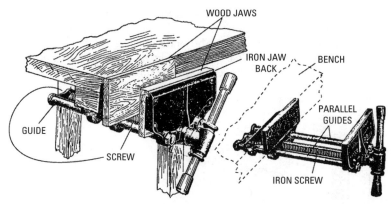

Figure 6-15 Bench vise with guide bars for keeping the jaws parallel. The inner jaw is fastened to the bench. It contains a fixed nut in which the screw rotates. This screw controls the outer jaw and draws it up tight against the workpiece. Two rods working through sleeves in the inner jaw keep the jaws parallel for all positions of the travel.

damage to the vise, support the end of the board. A system of supporting pegs carries the weight of the board and prevents stress on the vise. Drill a vertical row of holes in the leg at the right end of the bench (see Figure 6-13). Size the holes so a peg can be slipped easily from one hole to the next.

Tool Drop
A lower surface at the back edge of the bench top provides a place to keep tools that are currently being used (see Figure 6-13). This keeps them out of the way so large workpieces can lay flat across the main work surface. Although a few frequently used tools can be kept in the tool drop, clean it out often so it doesn't become permanent storage for supplies, sawdust, scraps, and other debris.

Storage
A system of drawers and shelves below the bench top can be designed to store tools and supplies (see Figure 6-13). This keeps them out of the way of current work but nearby when they are needed.

Bench Stop
This accessory is intended to prevent any lengthwise movement of the work while it is being tooled (that is, it prevents end-wise movement of a board while the board is being planed). As usually constructed for this purpose, a bench stop (see Figure 6-16) consists

of a metal casing that is designed to set in flush with the bench and has a horizontal toothed plate that works in vertical guides. A screw adjustment is provided so that the plate may be set flush with the top of the table (when not in use) or a little above so as to engage the end of the work and prevent endwise movement.

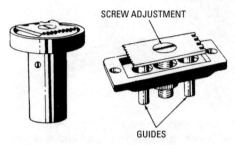

Figure 6-16 Round and rectangular forms of a bench stop. These adjust by a center screw from flush to as high as required for the work. The round bench stop is fitted by boring a hole the diameter of the stop with an expansion bit and a deeper center with the proper size of bit. The rectangular bench stop is shallow. It should be mortised in flush with the bench top.

Bench Hook
This is a movable stop. It can be used at right angles to the front of the bench. The hook serves many purposes for holding and putting work together. When you want to saw off a piece of stock, the bench hook is placed on the bench (see Figure 6-17). One shoulder is set

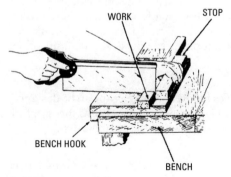

Figure 6-17 Typical bench hook and method of use when sawing to size with backsaw.

against the edge of the bench, and the upper shoulder serves as a stop for the work while sawing.

Summary

Holding tools are essential parts of workshop equipment. Many tooling operations require that the work be held rigid. The workbench can be considered a holding tool and should be anchored to the floor or wall to prevent movement.

Horses or trestles are a simple way to support work when large dimensions are being used.

An essential tool for any workbench is a vise. There are various types (such as woodworker's, bench, and sawhorse). The bench vise is not always a convenient tool to carry to a particular job, so in such cases clamps are used. Various types of clamps are available including C-clamps, miter clamps, and bar clamps.

A solid, steady workbench improves a worker's efficiency and quality of work. An effective bench is tailored to meet the needs of the work and the worker. Specialized bench designs suit the specific kinds of work done by the carpenter on the construction site and in the shop. Specific features, such as vises, stops, pegs, tool drops, and storage systems, turn a simple table into a specialized piece of equipment.

The bench hook is used for a variety of operations (such as odd sawing or chiseling), and serves to prevent the workbench from being marred by such operations.

Review Questions

1. What are some advantages of using a vise?
2. What is a sawhorse?
3. Why are clamps so important?
4. What is a chain clamp and when would it be used?
5. What is a hand screw?
6. What is another name for a sawhorse?
7. A holding tool is also called a _____.
8. Where did the C-clamp get its name?
9. What is the difference between a bar clamp and a C-clamp?
10. How wide do the jaws open on bench vises?
11. How does a workbench improve the accuracy and quality of work?

12. How does the design of a workbench affect the health of the person using it?

13. What does the weight of a bench have to do with where it is used and the kind of work done on it?

14. What is the purpose of a tool drop?

15. What feature of a bench is saved from damage by using support pegs when working on a long board?

16. What kind of bench would you use to cut the miter joints for interior window trim?

17. What device would be useful in cutting the window trim moldings to length?

18. What determines the height of a workbench?

19. For what is a bench hook used?

20. Why is the bench stop so important in bench work?

Chapter 7

Saws and Sawing

In almost all carpentry jobs, after the work has been laid out with guiding, marking, and measuring tools and supported or held in position by a holding tool, the first cutting operation will be performed by a toothed tool. The most important of these tools is the saw. Since sawing is hard work, the carpenter should know not only how to saw properly but also how to keep the saw in prime condition.

Saw Characteristics

Of course, a number of saws are available, and they can be characterized in a variety of ways, including the following:

- Type of blade
- Use
- Back reinforcement
- Others

However, the typical handsaw is the *crosscut* type, meaning that its blade is designed to cut wood across the grain. It is commonly 26 inches long.

Types

A handsaw that looks very much like the crosscut, differing in the design of the teeth, is the *ripsaw*. This saw is geared for cutting with the grain, or ripping a board (see Figure 7-1). Of course, most carpenters and handymen will use a power tool whenever possible for operations that involve extensive cutting.

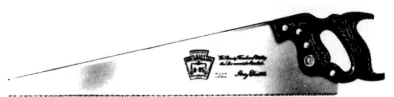

Figure 7-1 A typical handsaw used in various woodcutting operations. The coarseness or fineness of a saw is determined by the number of teeth per inch. A coarser saw, properly set, is preferred for fast work on soft and green wood, whereas a finer saw is suitable for smooth, accurate cutting and for dry, seasoned wood. Ripsaws commonly have 5½ to 6 teeth per inch, and crosscut saws have 7 or 8 teeth per inch.

Handsaws (crosscut or rip) will have a handle of wood or plastic and vary in length from 14 to 30 inches. The smaller sizes are called *panel saws*. Table 7-1 shows the lengths available in these saws.

Table 7-1 Saw Sizes

	Panel						Hand	Rip	
Size Inches	14	16	18	20	22	24	26	28	30

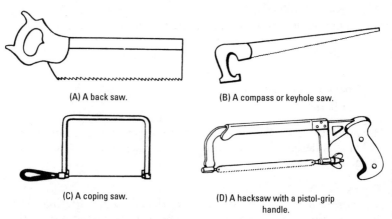

(A) A back saw.

(B) A compass or keyhole saw.

(C) A coping saw.

(D) A hacksaw with a pistol-grip handle.

Figure 7-2 Four popular types of saws.

Other saws are also very useful. They are designed to do jobs a regular handsaw cannot do, or cannot do easily. Following are other types of saws:

- *Backsaw* (see Figure 7-2A)—This is a thin crosscut saw with fine teeth that is stiffened by a thick steel web through the entire length of the blade along the back edge. One popular size has a 12-inch blade length with 14 teeth per inch. It is used for making joints and in fine woodworking operations where great accuracy is desired.

- *Compass (or keyhole) saw* (see Figure 7-2B)—This has a small narrow blade with a pistol-grip handle and is commonly used for cutting along circular curves or lines in fine or small work.

* *Coping saw* (see Figure 7-2C)—This is also used for cutting curves or circles in thin wood. It consists of a small narrow blade that is inserted in a sturdy metal frame in a manner similar to that used in the hacksaw.

* *Hacksaw* (see Figure 7-2D)—Although used primarily for cutting metal, this is a popular tool in any woodworking shop. There are two parts to a hacksaw: the frame and the blade. Common hacksaws may have either adjustable or solid frames, although the adjustable frame is generally preferred. Hacksaw blades of various types are inserted in these adjustable frames for different kinds of work; the blades vary in length from 8 to 16 inches. The blades are usually $1/2$ inch wide and have from 14 to 32 teeth per inch. A hole at each end of the blade allows the blade to be hooked to the frame. A wing nut on one end regulates the blade tension, thus permitting various types of cutting actions.

Still other saws with special uses are the *dovetail saw* (which is geared for cutting dovetail joints) and the *drywall saw* (which is used for cutting material of the same name). Of most recent vintage is the *rod saw*, which has a narrow blade coated with tungsten carbide particles and can be used to cut very hard materials, such as glass and ceramic tile.

Saw Teeth

The cutting edge of a handsaw is a series of little notches, all of the same size. On a crosscut saw, each side of the tooth is filed to a cutting edge like a little knife, as shown in Figure 7-3. On a ripsaw, each tooth is filed straight across to a sharp square edge like a little chisel (see Figure 7-4).

Set

The *set* of a saw is the distance that the teeth project beyond the surface of the blade. The teeth are set to prevent the saw from binding and the teeth from choking up with sawdust. In setting, the teeth are bent alternately, one to one side and the next to the other side, thus forming two parallel rows (or lines) along the edge.

Action of the Crosscut Saw

While each crosscut tooth resembles a little two-edged knife, it cuts quite differently. In early times, it was discovered that a knife blade must be free from nicks and notches to cut well. Then it could be pushed against a piece of wood, and a shaving could be whittled

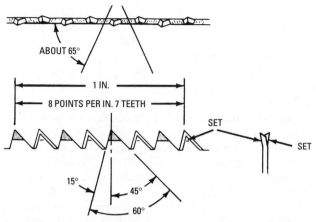

Figure 7-3 Side and tooth-edge views of a typical crosscut saw. This saw is used for cutting across the grain and has a different cutting action than that of the ripsaw. The crosscut saw cuts on both the forward and backward strokes.

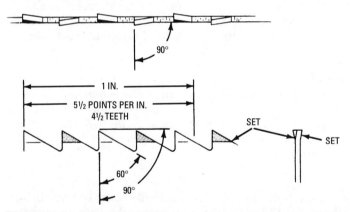

Figure 7-4 Side and tooth-edge views of a typical ripsaw. The ripsaw is used for cutting with the grain. Cutting is done only on the forward stroke.

off. At about the same time, it was noticed that if the nicked knife were drawn back and forth across the wood, it would tear the fibers apart, cutting a notch in the wood.

A modern saw's teeth are formed to cut a *kerf* (or notch) that is wide enough for the rest of the blade to pass freely. In this way, the saw can keep making the kerf deeper, eventually cutting the wood in two pieces.

The set of crosscut teeth makes them lie in two parallel rows. A needle will slide between them from one end of the saw to the other. When the saw is moved back and forth, the points (especially their forward edges) sever the fibers in two places, leaving a little triangular elevation that is crumbled off by friction as the saw passes through. New fibers are then attacked, and the saw drops deeper into the cut (see Figure 7-5).

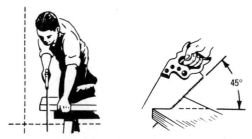

Figure 7-5   The approved method of crosscutting.

Action of the Ripsaw
The teeth of the ripsaw are a series of little chisels set in two parallel rows that overlap each other. At each stroke, the sharp edge chisels off a little from the end of the wood fibers (see Figure 7-6). The teeth are made strong with an acute cutting angle, but the steel is softer than that of a chisel to enable the teeth to be filed and set readily (see Figures 7-7 and 7-8).

Angles of Saw Teeth
The *face* of each crosscut tooth is slightly steeper than the back, thereby making an angle with the line of the teeth of approximately 66°. The compass teeth lean still further at an angle of 75°. The ripsaw face is at right angles (90°) to the line of the teeth. Its cutting edge is at right angles to the side of the blade. The angle of each tooth covers 60° (see Figure 7-9).

Coping Saw
With the coping (or fret) saw, the fork-shaped bench pin (made of wood) is temporarily clamped to the edge of the bench. A thin workpiece is held down on the top surface of the bench pin while the saw is worked up and down within the forked notch (see Figure 7-10).

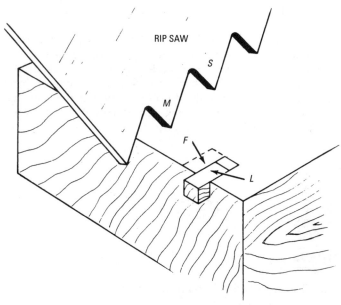

Figure 7-6 Ripsaw action. When the first tooth is thrust against the wood at an angle of approximately 45°, it chisels off and crowds out small particles of wood. Thus, tooth M will start the cut and take off piece L; tooth S will take off piece F, and so on.

Figure 7-7 Proper position for using the ripsaw.

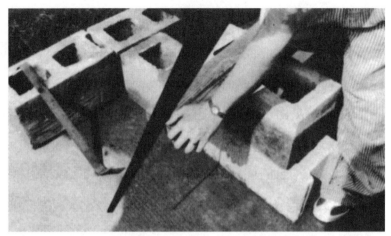

Figure 7-8 To saw, grasp the wood with the left hand, and guide the saw with the thumb. Hold the saw lightly, and do not press it into the wood. Simply move it back and forth, using long strokes.

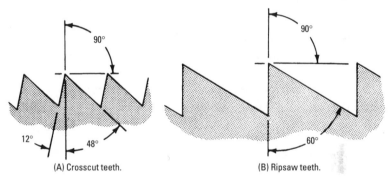

(A) Crosscut teeth.

(B) Ripsaw teeth.

Figure 7-9 Angular proportions for crosscut and ripsaw teeth.

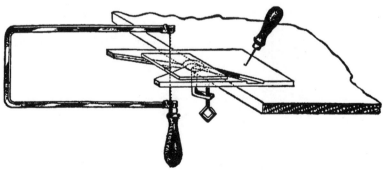

Figure 7-10 Method of using a coping saw or fret saw.

Summary

Various types of saws are used in carpentry work. Sawing is hard work. The carpenter should know how to saw properly and keep the saw in a sharp condition. There are many different kinds of saws, such as crosscut, rip, back, keyhole, coping, hacksaw, dovetail, and drywall.

Review Questions

1. How many teeth per inch are there in the average crosscut saw?
2. What is the difference between a crosscut saw and a ripsaw?
3. What is the difference between a coping saw and a keyhole saw?
4. Explain how a hacksaw functions.
5. Why must the saw teeth be set to project beyond the surface of the blade?
6. What is a saw kerf?
7. Where is the coping saw used?
8. Other than cutting wood, where is the compass or keyhole saw used?
9. Why is the backsaw ideal for use in a miter box?
10. What is the difference between rip and crosscut?

Chapter 8

Chisels, Hatchets, and Axes

The reason so many workers do not do quality finish woodwork is that they *do not keep their sharp-edged cutting tools sharp.* Not only should the edge be whetted as soon as any sign of dullness is observable, but also the tools should always be kept perfectly clean and free from rust. The tools to be considered here are what may be called *hand-guided* sharp-edged cutting tools (such as chisels and drawknives), as well as *striking* tools (such as hatchets) as opposed to *self-guided* tools (such as planes).

Chisels

In carpentry, the chisel is an indispensable tool. It is one of the tools most abused, because it is often used for prying open cases and even as a screw driver, although it is designed solely for cutting wood surfaces.

A chisel consists of a flat thick piece of steel with one end ground to an acute bevel to form a cutting edge and the other end provided with a wooden handle (see Figure 8-1).

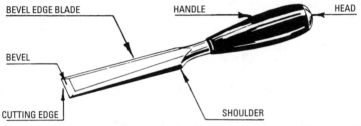

Figure 8-1   A typical general-purpose wood chisel with a handle of hard plastic, which is good at resisting breakage.

Chisels may be classified with respect to duty or service as follows:

- Paring
- Firmer
- Framing or mortise

Chisels may be classified with respect to the length of the blade as follows:

- Butt
- Pocket
- Mill

Chisels may be classified with respect to the edges of the blade as follows:

- Plain
- Bevel

Chisels may be classified with respect to the method of attaching the handle as follows:

- Tang
- Socket

Chisels may be classified with respect to the shape of the blade as follows:

- Flat
- Round (gouge)
- L (corner)

Paring Chisel

This is a light-duty tool for shaping and preparing relatively long planed surfaces, especially in the direction of the grain of the wood. The *paring chisel* (see Figure 8-2) is manipulated by a steady sustained pressure of the hand. It should not be driven by the blows of a hammer or other similar tool.

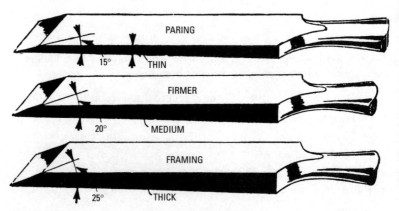

Figure 8-2 Paring chisel is for light duty; firmer is for medium duty. Chisels differ in the blade thickness. Paring is thin; firmer is thicker.

Firmer Chisel
The term *firmer* implies a more substantial tool than the paring chisel that is adapted to medium-duty work. The firmer chisel (see Figure 8-2) is a tool for general work and may be used for paring or light mortising. It is driven by hand pressure in paring and by blows from a mallet in mortising.

Framing or Mortise Chisel
This is a heavy-duty tool adapted to withstand severe strain (such as in framing, where deep cuts are necessary). In the best construction, an iron ring is fitted to the end of the handle to protect it from splitting, thus permitting the use of a heavy hammer in driving the tool into the wood.

Slick
Any chisel having a blade wider than 2 inches is called a *slick*. The regular sizes are $2\frac{1}{2}$, 3, $3\frac{1}{2}$, and 4 inches. Slicks are adapted for use on large surfaces where there is a great deal of material to be taken off, or where unusual power is required. They may be used to advantage in ship work in cutting down to a curve or bevel. They may be used either with a mallet or simply with the hands (see Figure 8-3).

Note
A hammer or other tool should not be used to drive a wood-handled chisel. Use only a mallet. Wood to wood in driving is the only satisfactory method for driving chisels.

Gouge
This is a chisel with a hollow-shaped blade for scooping or cutting round holes. There are two kinds of *gouge chisels*: the outside bevel and the inside bevel (see Figure 8-4). The outside bevel is more common.

Tang and Socket Chisels
According to the method by which the blade and handle are joined, chisels are called *tang* or *socket* (see Figure 8-5). The tang chisel has a projecting part (or tang) on the end of the blade that is inserted into a hole in the handle. The reverse method is employed in the socket chisel (that is, the end of the handle is inserted into a socket on the end of the blade).

Butt, Pocket, and Mill Chisels
This classification relates simply to the relative lengths of the blades. The regular lengths (see Figure 8-6) are approximately as

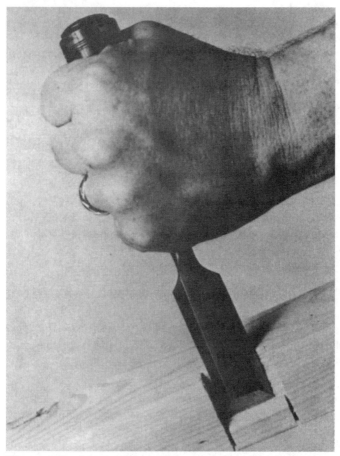

Figure 8-3   The butt chisel has a relatively short blade. Like all chisels, it is designed to be tapped on top of the handle, and it cuts with the flat side down. *(Courtesy of Stanley)*

follows: butt, $2^{1}/_{2}$ to $3^{1}/_{4}$ inches; pocket, 5 to 6 inches; and mill, 8 to 10 inches.

How to Select Chisels
A chisel should be flat on the back. An inferior chisel is ground off on the back near the cutting edge. In use, this results in a tendency to follow the grain of the wood, splitting it off unevenly because the user cannot properly control the tool. The flat back allows the

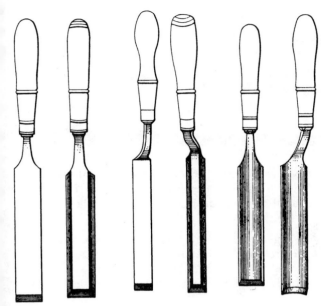

Figure 8-4   Several types of framing chisels and gouges.

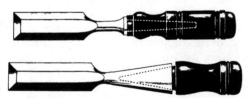

Figure 8-5   Typical tang (top) and butt socket chisels. The terms tang and socket are derived from the fact that the shank of the tang chisel has a point that is fitted into the handle. This point is called a tang, hence the name tang chisel. In the socket chisel, the shank of the chisel is made like a cup, or socket, with a handle fitted into it. Thus, the chisel is called a socket chisel.

chisel to take off the finest shaving, and where a thick cut is desired, it will not strike too deep. This is an important quality to look for in good chisels.

The best chisels are made of selected steel with the blade widening slightly toward the cutting edge. The blades are hardened, tempered, and carefully tested. The ferrule and blade of the socket chisel are so

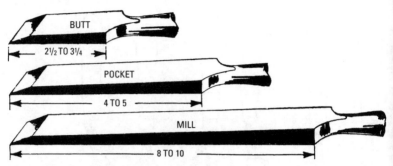

Figure 8-6 Wood chisels classified with respect to the length of the blade. The blade width may vary, depending on the type of work to be performed. The lengths given in the illustration are in inches.

carefully welded together that they practically form a single piece. Socket chisels are preferred to the tang type by most carpenters because they are stronger and the handles are less apt to split. Beveled edges are preferable to plain blades, because they have greater clearance.

The chisels commonly carried on construction jobs these days are quite short—only 3-inch or $3\frac{1}{2}$-inch beveled blades and metal-capped handles of a tough plastic that will take an amazing amount of abuse. The handles are cast and molded in place. They are designed for driving with a hammer. Such chisels are too short for a good job of paring or deep mortising, but they are serviceable for general heavy-duty use and for door-butt and similar mortising.

The butt chisel, because of its short blade, is adapted for close accurate work where not much power is required. It is particularly suited for putting on small hardware that does not necessarily require the use of a hammer. The butt chisel may be used almost like a jackknife, with the hand placed well down on the blade toward the cutting edge. The short blade and handle make it convenient for carrying in the pocket. Chisels are usually ground sharp and hand-honed and are ready for use when sold.

How to Care For and Use Chisels
To do satisfactory work with chisels, the following instructions should be carefully noted and followed:

- Do not drive the chisel too deep into the work. This requires extra pressure to dig out the chips.

- Do not use a firmer chisel for mortising heavy timber.
- Keep the tool bright and sharp at all times.
- Protect the cutting edge when not in use.
- Never use a chisel to open boxes, to cut metal, or as a screwdriver, putty knife or prying tool.

It should be noted that the chisel and the screwdriver are the tools most often misused by carpenters. They take more abuse than any other tool, not only in terms of care but also in terms of faulty use. For these reasons, the care practices suggested here should be followed closely, and the chisel should be carefully used (see Figures 8-7, 8-8, and 8-9).

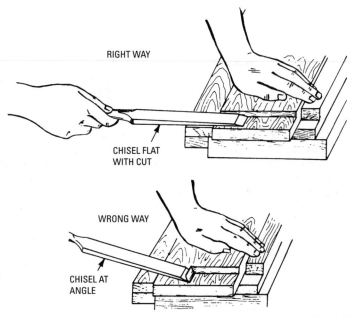

RIGHT WAY

CHISEL FLAT
WITH CUT

WRONG WAY

CHISEL AT
ANGLE

Figure 8-7 The right and wrong ways to use a paring chisel. The flat side of the chisel should face the cut and be parallel with it.

How to Sharpen Chisels
When honing a chisel, use a good grade of oilstone. Pour a few drops of machine oil on the stone. If you have no machine oil, lard can be used. The best results are obtained by using a *carborundum stone*. The carborundum cuts faster than most other abrasives, but

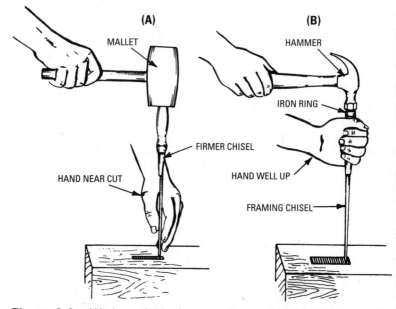

Figure 8-8 (A) A method of using a firmer chisel for light mortise work. Note the chisel is held close to end of blade. By holding it thus, it may be guided more accurately at the beginning of the cut. After starting a chisel accurately in a small mortise, the hand will naturally slip up toward the handle. The back of the chisel should be kept toward the end of the mortise toward which the chisel is approaching. Under no conditions should chisel cuts be made parallel with the grain until after the wood in the center of the mortise has all been cut out because the wood at the side may be split. (B) A method of using a framing chisel for cutting a large mortise. Note the chisel is held by hand near the handle end of the chisel. Here the chisel is shown driven by a hammer instead of a mallet. This is permissible with a framing chisel when the handle is properly reinforced by an iron ring as shown.

the edge will not be as smooth and keen as when a natural oilstone is used.

Hold the chisel in the right hand, and grasp the edges of the stone with the fingers of the left hand to keep it from slipping. A better method is to place the stone on a bench and block it so it cannot move. Both hands will then be free to use for honing. In this case, grasp the chisel in the right hand where the shoulder joins the socket.

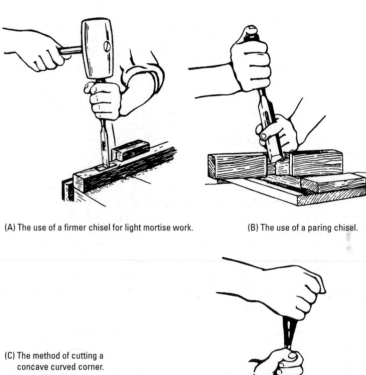

(A) The use of a firmer chisel for light mortise work.

(B) The use of a paring chisel.

(C) The method of cutting a concave curved corner.

Figure 8-9 Using chisels for various types of work.

Place the middle and forefinger on the blade near the cutting edge. Rub the chisel on the oilstone away from you. Be careful to keep the original bevel.

Never sharpen the chisel on the back or flat side. This should be kept perfectly flat. For paring, the taper should be long and thin—approximately 15° (see Figure 8-2). The longer the bevel on the cutting edge, the easier the chisel will work and the easier it is to hone. A firmer chisel should be ground at an angle of not less than

20°. An angle of 25° is recommended for a framing chisel. When honing a chisel, the taper should be carefully maintained, and, unless the back is kept flat, it will be impossible to work to a straight line. Bevel-edged chisels are more easily sharpened than the plain-edged type because there is not as much steel to remove.

If the chisel is badly nicked, it will have to be ground before honing. Not many quality chisels can be filed. Do not overheat or damage the temper of the chisel, and be sure to keep the original taper of the bevel. After grinding, hone the chisel on an oilstone as detailed earlier.

Drawknife

This tool consists of a large sharp-edged blade with a handle at each end, usually at right angles to the blade (see Figure 8-10). It is used for trimming wood by drawing the blade toward the user. When the blade is sharp, and some degree of force is applied, it does its work quickly and efficiently.

The tool was formerly used for the rapid reduction of stock to an approximate size, an operation that is now performed by sawing

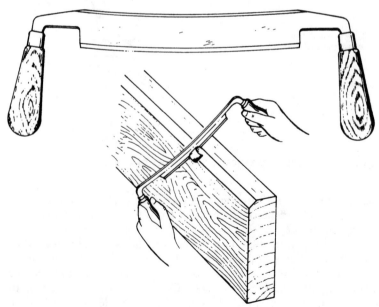

Figure 8-10   A typical drawknife and its use.

or planing machines. The drawknife is, however, quite effective on narrow surfaces that must be considerably reduced. Drawknives are made with cross-sections of various shapes, thereby adapting them for a variety of jobs (see Figure 8-11).

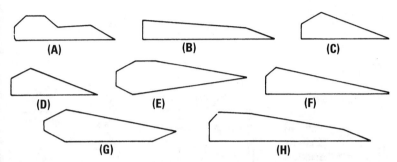

Figure 8-11   Cross-sections of various drawknife blades. (A) Carpenters' razor blade; (B) Carpenters' light blade; (C) Carriage makers' narrow blade; (D) Coach makers' razor blade; (E) Wagon makers' heavy blade; (F) Shingle shave, heavy blade; (G) Saddle tree shave, heavy blade; (H) Spar shave, heavy blade.

Hatchets and Axes

The hatchet is a general utility tool familiar to all. In framing timber, it can be used as a hammer, for sharpening stakes, for cutting down timber to rough size, or for splitting wood. For lathing, it is a combination hammer and cutting-off tool. It serves a similar purpose in shingling (Figures 8-12 and 8-13).

Broad Hatchet or Hand Axe

This is simply a large hatchet with a broad cutting edge. Ordinarily it is grasped with the right hand at a distance of about one-third from the end of the handle, but the position of the hand will be regulated in great measure by the material to be cut (that is, by the intensity of the blow). Thus, to deliver a heavy blow, the handle is grasped close to the end, and for a light blow, nearer the head (see Figure 8-14). Figure 8-15 shows the proper beveling when sharpening a hatchet or ax.

Axe

This tool is similar to the hand axe but of larger size with a longer handle intended for heavy cutting using both hands (see Figure 8-16).

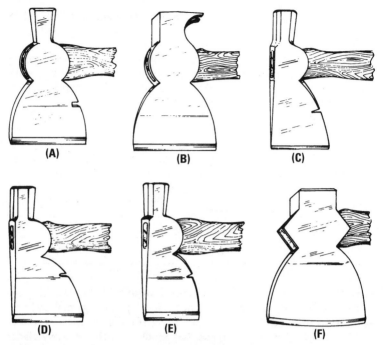

Figure 8-12 Various types of hatchets. (A) Shingling; (B) Claw; (C) Barrel; (D) Half; (E) Lath; (F) Broad (hand axe).

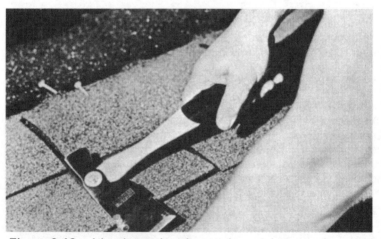

Figure 8-13 A hatchet such as the one shown is being used to establish length of shingle to the weather—how much is exposed. The gage pin in the hatchet is adjustable to weather length desired. Striking face is beveled and tempered for driving nails. *(Courtesy of Vaughn & Bushnell)*

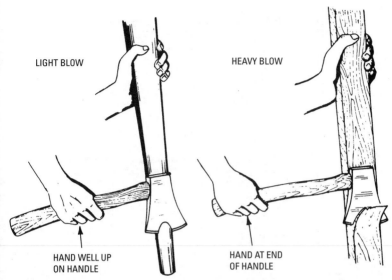

LIGHT BLOW

HEAVY BLOW

HAND WELL UP
ON HANDLE

HAND AT END
OF HANDLE

Figure 8-14 Grasp the handle of the hand axe approximately halfway between the ends to strike a light blow and at the end of the handle to obtain the necessary swing for a heavy blow.

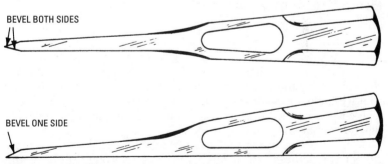

BEVEL BOTH SIDES

BEVEL ONE SIDE

Figure 8-15 When sharpening a hatchet or axe, bevel both sides of the cutting edge for general use. For hewing to a line, bevel only one side.

Figure 8-16 A typical hatchet and an axe.

Adze

Briefly, an adze is a form of hatchet in which the edge of the blade is at right angles to the handle. The blade is curved or arched (toward the end of the handle), thus permitting an advantageous stroke of the tool while the operator is standing over the work (see Figure 8-17). The edge is beveled on the inside only, the handle being removed when necessary to grind the tool.

In house carpentry the adze is not so much used as when more timber was hand-hewn. It has many uses for rough dressing and shaping preparatory to finishing tools (such as in log cabin work). An important use for the adze is in ship carpentry.

Amateurs are advised to give the adze a wide berth. It is a dangerous tool. Even ship carpenters often inflict severe wounds on their feet with this tool. The edge of the adze must be very sharp and in perfect condition (no nicks).

Figure 8-17 Method of using the adze illustrating its dangerous character. Evidently, since the work is usually held by the foot, there is danger of a misdirected blow cutting the foot, especially as the tool to be of any use must have a very keen edge.

In sharpening an adze (or an axe), the tool should be traversed across the face of the grindstone, holding it at the proper angle until all the nicks have been taken out. Then, to secure a keen edge, rub with a slip of stone. It is important in sharpening an adze to bevel only on the inside.

Summary

Poor results in woodworking are often caused by using dull cutting tools (that is, neglecting to keep tools sharp). The cutting edges should be whetted or honed as soon as any sign of dullness is observed. The tools should always be kept perfectly clean and free from rust. Sharp-edged cutting tools may be divided into several classes (such as chisels, planes, drawknives, hatchets, and axes).

When honing a chisel, use a good grade of oilstone. Pour a few drops of machine oil on the stone and rub the chisel away from you, being careful to keep the original bevel. Never sharpen a chisel on the back or flat side. This should be kept perfectly flat. The best chisels are made of selected steel with the blade widening slightly toward the cutting edge.

A hatchet can be used as a hammer, for sharpening stakes, cutting timber to a rough size, or splitting wood. The hatchet is also used in roof shingling.

There are various types of hatchets (such as shingling, claw, barrel, half, and lath). The cutting edge must be free of nicks to be a useful tool.

Review Questions

1. What is the difference between a tang chisel and a socket chisel?
2. What is a paring chisel?
3. What is a drawknife?
4. What are the three basic blade lengths for chisels?
5. What is a corner chisel?
6. What is the difference between a hatchet and an axe?
7. What is an adze?
8. In what type of work would a hatchet be used?
9. How are hatchets used in framing work?
10. Why should cutting edges be kept razor-sharp?

Chapter 9

Planes, Scrapers, Files, and Sandpaper

The tools under this classification are those sharp-edged cutting tools in which the cutting edge is guided by the contact of the tool with the work, instead of being guided by hand. For example, consider a plane as distinguished from a chisel. The plane, being positively guided, gives a smooth cut in contrast with the rough cut obtained by the hand-guided chisel—hence the term *smoothing tools*. These tools are essentially chisels set in appropriate frames, so that the contact of the frame with the work during the movement of the tool will give a positive guide to the cutting edge, thus resulting in a smooth cut.

Spoke-Shave

This tool resembles a modified drawknife whose blade is set in a boxlike frame that forms a positive guide. The blade is adjustable like a plane to govern the thickness of the cut. *Spoke-shaves* may be made of wood or metal, and they have obtained their name from the fact that they were once used in the making of wagon spokes. They are very useful in smoothing curved edges and to round irregular surfaces (see Figure 9-1). Spoke-shaves are made with cutters of various shapes (straight, hollow, round, angular, and so on).

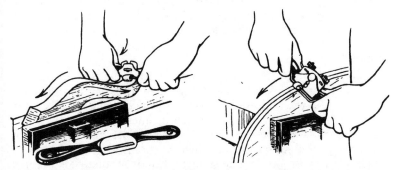

Figure 9-1 A method of using the spoke-shave. The spoke-shave cuts when pushed away from the user, as indicated by the arrow. Be careful to work in the direction that the tool cuts without tearing the grain. The spoke-shave is also used to chamfer and cut edges.

The flat-bottom spoke-shave is used on convex and concave surfaces where the curves have a long sweep. It is also used to chamfer or round edges. The hollow-bottom (or concave-bottom) spoke-shave is used for rounding edges that have small convex sweeps, and the convex-bottom spoke-shave is employed to cut concave curved edges that have small sweeps.

Spoke-shave cutters may be sharpened by removing the blade from the stock and rubbing it on the inside with a flat slip of oilstone. Lightly rub the outside of the blade on an ordinary oilstone (see Figure 9-2). To hold the small blade more firmly, place it into a saw kerf made across the end of a small, flat piece of wood, with the edge of the blade projecting beyond the wood. The piece of wood should be beveled to allow the blade to lie on the stone at the proper angle. It may then be sharpened like a plane iron.

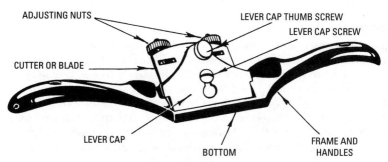

Figure 9-2 A typical spoke-shave. This is a lightweight, handy tool for use on concave, curved edges that have large or small sweeps. The wing-nut adjustment is on the cutter cap.

Planes

A *plane* is a tool for smoothing boards or other surfaces of wood. It consists of a stock (usually made of wood or iron, or a combination of both), from the underside (or face) of which slightly projects the cutting edge. The cutting edge, which inclines backward, is called the *plane iron*. An aperture in the front provides for the escape of the shavings that are produced when the tool is in action.

The plane is essentially a finishing tool, and, although it is adapted for use in bringing wood surfaces to the desired thickness, it will produce this result only gradually as compared to a chisel or hatchet. For this reason, it is normally the last tool to be used in finishing a wood surface.

There are many planes to meet varied requirements (see Figure 9-3). Useful planes include the following:

- Jack plane
- Fore plane
- Jointer plane
- Smoothing plane
- Block plane
- Molding plane
- Rabbet plane
- Grooving plane
- Router plane

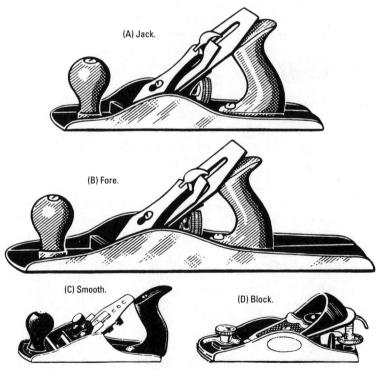

(A) Jack.

(B) Fore.

(C) Smooth.

(D) Block.

Figure 9-3 Various types of planes used by woodworkers.

Jack Plane

This plane is intended for heavy, rough work. It is generally the first plane used in preparing the wood. Its purpose is to remove irregularities left by the saw and to produce a smooth surface. The *jack plane* is long and heavy enough to make it a powerful tool, so it will remove a considerable chip with each cut. The cutting edge of the plane iron is ground slightly rounded. This form is best adapted for roughing. If properly sharpened, the jack plane may be used as a smoothing plane or as a jointer on small work because it is capable of doing just as good of work.

Fore Plane

This plane is designed for the same purpose as the jointer plane (that is, to straighten and smooth the rather rough and irregular cut of the jack plane). Since the *fore plane* is shorter than the jointer (usually 18 inches in length), it is easier to handle (especially for a journeyman carpenter), and it may be used as a jack plane. If a carpenter does not have both a jack plane and a jointer, a fore plane can serve for both, although it will not give as good service as either of the other two in the work for which they are adapted. The plane iron of the fore plane is sharpened to a straight line and is set for a finer cut than that of the jack plane.

Jointer Plane

The great length and weight of this plane keeps the cutter from tearing the wood, and, with the cutter set for a fine cut, it is the plane to use for obtaining the smoothest finishes. These planes will true up somewhat better than other types of planes.

In this country, the word *jointer* is applied to planes that range in size from 22 to 30 inches. The length of the plane determines the straightness of the cut. Thus, a smoothing plane (because of its short length) will follow the irregularities of an uneven surface, taking its shavings without interruption. A fore or jointer plane similarly used will first touch only the high spots, progressively lengthening the cuts until, on reaching the lowest spots, a continuous shaving will be taken. The final cut will approach a true surface depending on the length of the tool and the length of the irregularities or undulations that were there to begin with. The cutting edge of a jointer plane is ground straight and is set for a fine cut.

Smoothing Plane

The small length of this plane (usually about 8 inches) adapts it for finishing uneven surfaces. Because of its small size, it will find its way into minor depressions of the wood without taking off much

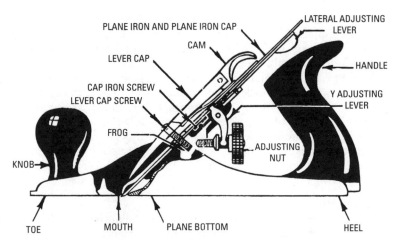

Figure 9-4   A typical smoothing plane and its component parts.

material. In this respect, the *smoothing plane* differs from the jointer plane (see Figure 9-4). Although both are finishing planes, the jointer plane is used for finer work.

Block Plane

This type of plane (see Figure 9-5) is the smallest plane made (4 to 7 inches in length). It was designed to meet the demand for a plane

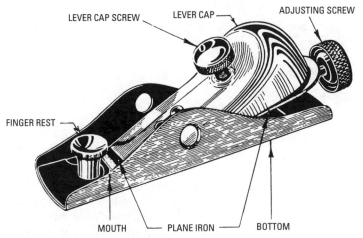

Figure 9-5   A typical block plane and its component parts.

that may easily be held in one hand while planing across the grain. The block plane is used almost exclusively for planing across the grain. Therefore, no cap iron is necessary to break the shavings because they are only chips.

The bevel of the plane iron is turned up instead of down. Because of its size, the block plane is usually operated with one hand, with the work held by the other hand. Therefore, as distinguished from this method of use, other planes are called *bench planes*. The angle of the plane iron for block planes is much smaller than that for bench planes. This angle is 20° for softwoods and 12° for hardwoods. Planes with the iron set at 12° are called *low angle block planes*.

Rabbet Plane

In this type of plane, the plane iron projects slightly from the side, as well as from the bottom of the plane (see Figure 9-6). There are various forms of rabbet planes available, each suitable for different types of cuts. With a tool of this type, the edge of a board can be cut so as to leave a rabbet or sinking (like a step) along its length to fit over and into a similar indentation cut in the edge of another board. Rabbet planes are adapted to cut with or across the grain according to the setting of the iron.

Figure 9-6 A typical rabbet plane.

Surform

The *Surform* (see Figure 9-7) from Stanley Tools is also in the plane family. Like other tools in this line, it has a cheese-grater-like cutting surface and makes short work of wood, soft metals, and other materials. The tool is pushed along like any plane to do its cutting.

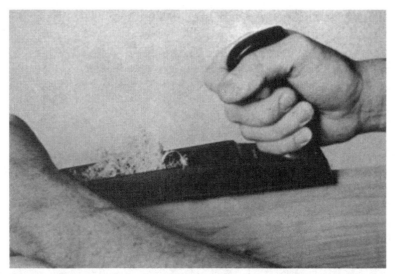

Figure 9-7 A Surform plane from Stanley. It removes a large amount of material quickly. *(Courtesy of Stanley)*

Grooving Plane

The *grooving plane* (sometimes called a *trenching plane*) is used for cutting grooves across the grain. It has a rabbeted sole. The cutters are in the tongue portion that is usually ½ inch deep and varies from ¼ to 1⅛ inches. A screw stop adjusts the depth of the cut, and a double-toothed cutter separates the fibers in front of the iron (see Figure 9-8).

Figure 9-8 A typical grooving, or plow, plane, especially designed for weatherstrip grooving.

Router

Planes of this type are used for surfacing the bottoms of grooves or the like that are parallel with the general surface of the wood. The closed-throat type is the ordinary form of *router*. The open-throat type is an improved design, giving more freedom for chips and a better view of the work and cutter (Figure 9-9). The open-throat router has an attachment for regulating the thickness of the chip and a second attachment for closing the throat for use on narrow surfaces. The bottoms of both styles are designed so that an extra wooden bottom of any size desired can be screwed on, thereby enabling the user to rout large openings.

Figure 9-9 Typical router planes. This plane is used for surfacing the bottom of grooves or other depressions parallel to the work.

Plane Irons or Cutters
The plane iron that does the cutting is similar to a chisel, but differs in that its sides are parallel and the thickness is less than that of the chisel blade (see Figure 9-10).

Plane irons are classed with respect to thickness as follows:

• Heavy
• Thin

Plane irons are classed with respect to the shape of the cutting edge as follows:

• Curved
• Straight (square)

Plane irons are classed with respect to provisions for breaking the chips as follows:

• Single (see Figure 9-11)
• Double (see Figure 9-12)

Figure 9-10   The iron on a plane can be easily removed for sharpening or manipulated for setting. *(Courtesy of Stanley)*

Heavy plane irons are usually No. 12 gage, whereas the medium or thin plane irons are usually No. 14 gage. The heavy plane iron offsets the tendency found in spring-cap planes to vibrate. The additional weight helps avoid chattering.

The thin plane iron is normally satisfactory when the plane is properly constructed (that is, firm support is given the cutter over a considerable portion of its length). It also has the slight advantage of requiring less grinding.

For the first (or roughing) cut with the jack plane, the cutting edge is ground slightly curved (convex) (see Figure 9-13) because, since

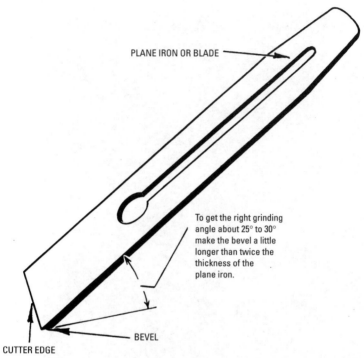

PLANE IRON OR BLADE

To get the right grinding angle about 25° to 30° make the bevel a little longer than twice the thickness of the plane iron.

BEVEL

CUTTER EDGE

Figure 9-11 A single plane iron.

it is used for heavy work, it removes thick shavings. If the cutter were ground straight, the plane would cut a rectangular channel from which the wood would have to be torn as well as cut (see Figure 9-14A). Moreover, such a shaving would probably stick fast in the throat of the plane or require undue force to push the plane. Compare this with the shaving taken from the fully curved cutting edge of the jackplane (see Figure 9-14B).

When a full set of planes is available, the fore plane should have some curvature to the cutting edge. In this case, the process of transforming the grooved surface produced by the jack plane to a flat surface is accomplished in three operations, using the jack, fore, and jointer planes (see Figure 9-15). The cutting edge of the jointer and smoothing plane irons are made straight with rounded corners (see Figure 9-13). Because this type of plane iron makes an extremely fine cut, the groove caused by the removal of so delicate a shaving is

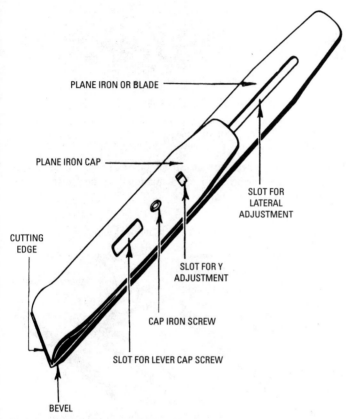

Figure 9-12 A double plane iron.

sufficiently blended with the general work by the rounded corners of the iron.

Bevel of the Cutting Edge
Many of the complaints concerning poorly cutting plane irons are because of improper plane-iron grinding. The bevel should always be ground at an angle of 25° to 30°. This means it must be twice as long as the cutter is thick. If the bevel is too long, the plane will jump and chatter. If it is too short, it will not cut. It is a good rule, perhaps, to have a long thin bevel for softwood and a 25° bevel for the hardwoods, although cross-grained timber requires a short bevel.

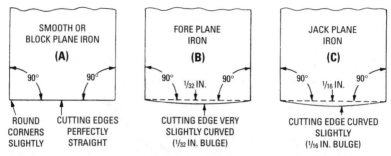

Figure 9-13 Cutting edges for common plane irons. These edges should be straight on smooth and block plane irons and just slightly curved on jack and fore plane irons.

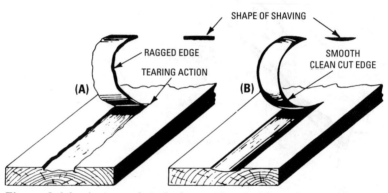

Figure 9-14 Actions of a jack plane with a straight and a curved cutter.

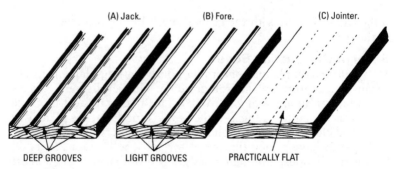

Figure 9-15 Appearance of a board surface after having been planed with jack, fore, and jointer planes.

Double Irons
The term *double iron* means a plane iron equipped with a supplementary iron called a *cap*. The object of the cap is to break the shaving as soon as possible after it is cut. Figure 9-16B shows the action of the cap.

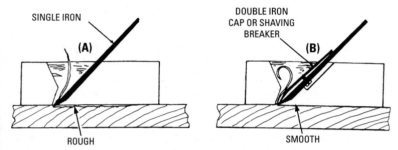

Figure 9-16 Action of single and double plane irons. The single iron cuts satisfactorily only when the grain is favorable. But when the grain varies from the line of cut, the shavings run up the iron, leaving a rather rough surface.

The cap is attached to the cutting iron by tightening a screw that passes through a slot in the cutter. The distance the cap is placed from the edge of the plane iron varies with the thickness of the shaving. Allow $^1/_{32}$ inch for a smooth or fore plane and approximately $^1/_8$ inch for a jack plane.

Plane Mouth
This is the rectangular opening in the face of the plane through which the cutter projects and, in operation, through which the shavings pass. The width of the mouth has an important bearing on the proper working of the plane. That portion of the plane face in front of the mouth prevents the wood from rising in the form of a shaving before it reaches the mouth. If there were no face in front of the cutter (as in the case of a bull-nose plane), there would be nothing to hold down the wood in advance of the cutter, and the shaving would not be broken. In obstinate grain, the work will be rougher, and a splitting instead of a cutting action may result (see Figure 9-17). Accordingly, the wider the mouth, the less frequently the shaving will be broken and, in tough grain, the rougher the work.

How to Use a Plane
To obtain satisfactory results with planes, it is necessary to know not only the proper method of handling the tool in planing, but

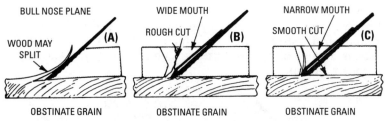

BULL NOSE PLANE   WIDE MOUTH   NARROW MOUTH

ROUGH CUT   SMOOTH CUT

WOOD MAY SPLIT   **(A)**   **(B)**   **(C)**

OBSTINATE GRAIN   OBSTINATE GRAIN   OBSTINATE GRAIN

Figure 9-17   Influence of mouth width. (A) The bull-nose plane may be regarded as an ordinary plane with a mouth of infinite width. Since there is nothing in front of the cutter to hold down the wood, a splitting action is possible with obstinate grain. (B) and (C) show the results obtained with wide- and narrow-mouth planes.

also how to put it into good working condition. The carpenter must know the following:

- How to sharpen the cutter
- How to adjust the cutter
- How to plane

Sharpening Planes

This involves two operations: grinding and whetting. When grinding, the cutter must be ground perfectly square (that is, the cutting edge must be at right angles to the side). Enough metal must be removed to remove all nicks in the cutting edge. Before grinding, loosen the cap and set it back approximately $1/8$ inch from the edge. It will serve as a guide by which to square the edge.

The cutter should be held firmly on the grinding wheel at the proper angle and should be moved continually from side to side to prevent wearing the stone out of true. Grind on the bevel side only. The bevel angle should be approximately 30°. The angle is attained when the length of the bevel is twice the thickness of the cutter (see Figure 9-18). The edge should be ground to one of the forms shown in Figure 9-13, depending on the type of plane and the requirements of the work to be done.

After grinding, the cutter will have a wire edge (that is, the coarse grit of the grinder will always leave the edge comparatively coarse or rough, and the edge will not be as keen as it should be to cut smoothly). This wire edge is removed by the aid of an oilstone (see Figure 9-19).

In the case of a double iron, the cap should be kept with a fine (but not a cutting) edge. The cap must be made to fit the face of the

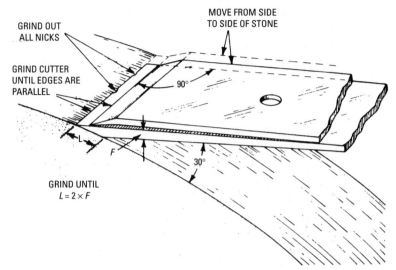

Figure 9-18 The proper position of the plane iron on the grinding wheel. Note the conditions that must be fulfilled to grind properly.

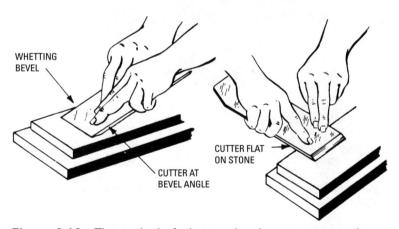

Figure 9-19 The method of whetting the plane iron on an oilstone after grinding. Grasp the plane iron firmly in the right hand with the palm down, pressing down with the left hand near the cutting end to provide rubbing pressure. Rub back and forth along the length of the stone. After whetting the bevel side, turn the plane iron over and hold it perfectly flat on the stone. Give it two or three strokes to remove any wire edge.

cutter accurately. If it does not fit precisely, the plane will quickly choke with shavings because of the shavings driven between the two irons. This is an extremely important point, and it should be noted that even a minute opening between the plane irons will allow the shavings to drive in and choke the plane.

A honing guide makes the sharpening job easier. The plane iron can be inserted in the device and set for the proper angle, then the blade can be drawn back across the stone with no errors possible. The device can also be used for sharpening spoke-shave and wood-chisel blades (see Figure 9-20).

Figure 9-20 A honing guide makes sharpening plane irons easier.
(Courtesy of Stanley)

Adjusting the Cutter

After sharpening the cap of a double iron, position the screw in the slot. Tighten the screw lightly on the cap to within $1/4$ inch of the cutting edge, then tighten the screw. Finish the setting by driving the cap up to its final position, tapping lightly on the setscrew.

The *set* of the iron is the amount of cutter face exposed below the edge of the cap. The plane iron is set coarse or fine according to the amount of cutter face exposed. The set, therefore, regulates the thickness of the shavings. It is varied according to the nature and kind of wood to be planed. For softwoods, the set should be $1/4$ inch for the jackplane, $1/16$ inch for the jointer plane, and $1/32$ inch for the smoothing plane. If the wood is hard or cross-grained, allow approximately one-half of these settings.

How to Plane
Satisfactory results in the use of a plane depend largely on the plane being in perfect condition and properly adjusted with respect to set and depth of cut to suit the kind of wood to be planed. The first thing to learn is the correct way of holding the plane. The plane should not drop over the end of the board at either end of the stroke. Before planing, examine the board with respect to the grain, and turn the material to take advantage of the grain. On the return stroke, lift the back of the plane slightly so that the cutter does not rub against the wood, thus preventing the cutter from quickly being dulled.

When planing a narrow surface, let the fingers project below the plane, and use the side of the board as a guide to keep the plane on the work. If the plane chokes with shavings, look for and repair the cause instead of just removing them. Remove the iron and carefully examine the edge of the cap. This must be a perfect fit or there will be continual trouble.

To plane a long surface (such as a long board), begin at the right-hand end. Take a few strokes, then step forward and take the same number of strokes, progressing this way until the entire surface is passed over.

When cutting across the grain with a block plane, the cut should not be taken entirely across, but the plane should be lifted before the cutter reaches the edge of the board. If this precaution is not taken, the wood will split at the edge. After taking a few strokes, reverse the board and continue as directed (see Figure 9-21).

Take care when planing the edges of a piece to avoid splitting off ends. It's safest to work from the edge in toward the center of the work (see Figure 9-22). If you are planing all four edges of a board, it may be practical to plane in a clockwise or counterclockwise fashion, so that if the planing chips adjacent edges you can shave these off on subsequent passes.

Scrapers
The term scraper usually signifies a piece of steel plate of about the thickness and hardness of a saw (see Figure 9-23). Following are

Figure 9-21 The method of using the block plane across the grain. The cut must not be taken across the entire length of the board to prevent the board from splitting. Lift the plane before the cutter runs off the edge. Take several strokes with the board in position MS. Then reverse the board to position SM and continue planing.

Figure 9-22 Planing near an edge, work from the edge in toward the center. *(Courtesy of The American Plywood Assn.)*

types of scrapers:

- Unmounted
- Handle scraper
- Scraper-plane

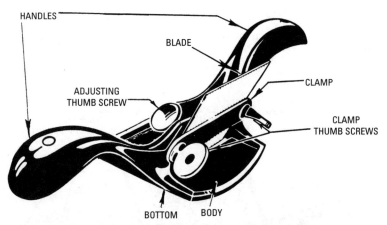

Figure 9-23   A typical cabinet scraper. In operation, the blade springs backward, thereby opening the mouth and allowing the shavings to pass through. As soon as the working pressure is released, the blade springs back to its normal position.

The *unmounted scraper* is simply a rectangular steel blade, whose cutting edges are formed by a surface at right angles to the sides (see Figure 9-24). Quicker cutting is secured by having the cutting edge more acute, but more labor is required to keep it sharp. The cutting edge is sharpened by filing or grinding.

For smoother work, the roughness of the edge may be removed by an oilstone, but the rougher edge will cut faster.

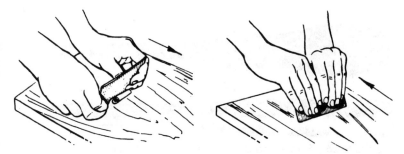

Figure 9-24   A hand scraper and its method of use. In operation, the hand scraper is pushed or pulled, thus removing surface irregularities in the work. When properly used, it provides the work with a smooth and glossy finish.

In cutting, the scraper is inclined slightly forward and is more conveniently held when provided with a handle or mounted as a plane (see Figure 9-25).

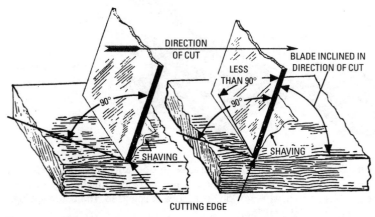

Figure 9-25 Scraper with normal (90°) cutting edge and acute (less than 90°) cutting edge, showing position of scraper in cutting.

A good method of sharpening the scraper to avoid too frequent use of the oilstone is as follows:

- File the cutter to a keen edge, removing wire edge with a coarse, medium oilstone.
- Holding the burnisher in both hands, turn the edge (see Figure 9-26).

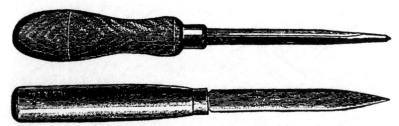

Figure 9-26 Scraper steels or burnishers, oval form and round form. These tools are used to turn the cutting edge of a scraper after filing and honing. Turning means pushing over the particles of steel that form the corner so that they will form a wire edge that will stand at an angle with the sides of the scraper.

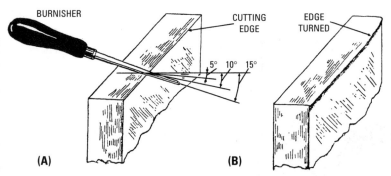

Figure 9-27 Application of burnisher in turning the edge of a scraper after filing and honing. (A) shows angles at that burnisher should be held. The edge is usually turned in two or three strokes with burnisher at 5°, 10°, and finally at 15° as shown. (B) shows appearance of turned edge (greatly enlarged).

- Begin with light pressure and hold the steel at nearly the same angle as the file was held in filing (see Figures 9-27 and 9-28).
- Bear down harder for each successive stroke, and let the tool come a little nearer level each time, finishing with tool at angle of about 60° from the face of the blade.
- Be sure that the steel never comes down squarely on the fine edge, for that will ruin it.
- Keep the edge a little ahead of the face of the cutter. The object is to get a sharp hook-edge.

Files and Rasps

By definition, a file is a steel instrument having its surface covered with sharp-edge furrows or teeth that is used for abrading or smoothing other substances (such as metal and wood). A *rasp* is a coarse file and differs from the ordinary file in that its teeth consist of projecting points instead of grooves cut across the face of the file.

Files are used for many purposes in woodworking. Figure 9-29 shows a variety of them. The *taper file* is adapted for sharpening hand, pruning, and bucksaws. The teeth of the *mill file* leave a smooth surface. They are particularly adapted to filing and sharpening mill saws and mowing- and reaping-machine cutters. Rasps are generally used for cutting away or smoothing wood or for finishing off the rough edge left in a circular hole that has been cut with the

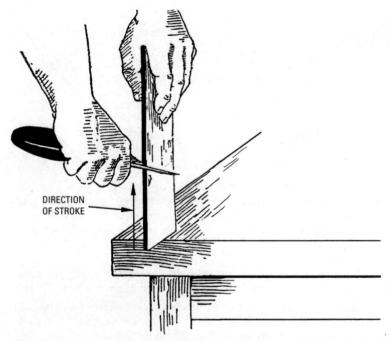

Figure 9-28 Turning edge with burnisher. Note angle at which the tool is held. The stroke is made from bottom up as indicated by the arrow. Slightly lubricate burnisher to assist it in sliding over the edge of the scraper without scratching.

keyhole saw. The ordinary *wood rasp* is rougher or coarser than that used by cabinetmakers. Wood files are usually tempered to cut lead or soft brass.

In using large rasps or files, whether for wood or metal, the work should be held in the vise or otherwise firmly fixed, because it is desirable to use both hands when possible. The handle of the tool should be grasped by one hand while the other hand is pressed, but not too heavily, on the end or near the end of the blade to lend weight to the tool and add to its powers of abrasion.

Sandpaper

While not a tool in the strictest sense, sandpaper most certainly is a cutter, and knowledge of it is well advised for anyone working with wood.

In reality, sand is never used in making sandpaper, and the backing material is often not paper. It is paper, paper/cloth combinations, and

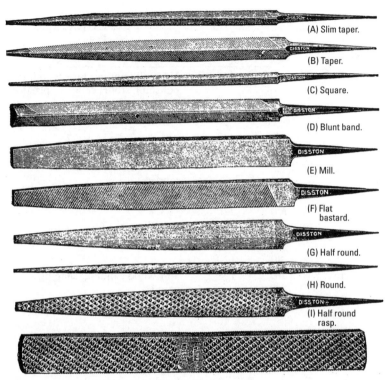

Figure 9-29   Various types of files and rasps.

various fibers. Properly speaking, the name for sandpaper is *coated abrasive*.

Sandpaper is characterized in a number of ways: by the so-called retail system (where there are word descriptions of its uses), by numbers, by the industrial system (which is widely used), and by the old system (which is also numbers). Table 9-1 shows various types of sandpaper.

The grit on sandpaper may be composed of flint, emery, garnet, aluminum oxide, or silicon carbide. Flint is cheap but ineffective over a long period. Use it for removing paint. Garnet cuts better than flint and lasts longer. Paper with this grit is generally useful all-around. Emery is a black abrasive that has been superseded by other abrasives but lasts, it seems, simply because the newer materials are not well known.

Aluminum oxide lasts longer and cuts much better than the other three papers. Today it is regarded as the best all-around sandpaper to use.

Table 9-1 Designations and Uses of Sandpaper

Retail	Old System	Industrial System	Uses
Super fine or extra fine	—	600	Normally not used for wood. For polishing stone, ceramic, plastics.
	—	500	
	10/0	400	
	9/0	320	
Very fine	8/0	280	For polishing wood finishes between coats, usually wet sanded.
	7/0	240	
	6/0	220	
Fine	5/0	180	Fine sanding of bare wood.
	4/0	150	
	3/0	120	
Medium	2/0	100	Soft wood sanding; shaping wood.
	1/0 or 0	60	
	$\frac{1}{2}$	40	
Coarse	1	50	First, rough sanding. Removing paint.
	$1\frac{1}{2}$	60	
Very coarse	2	36	Used for sanding floors.
	$2\frac{1}{2}$	30	
	3	24	
Extra coarse	$3\frac{1}{2}$	20	Also used on floors, particularly where floor is painted.
	4	16	
	$4\frac{1}{2}$	12	

Silicon carbide is blue-black in color and is the hardest abrasive of all. It is used on metal, glass, and similarly hard substances.

Summary

The woodworking plane consists essentially of a smooth-soled stock of wood or iron. A cutting edge projects from the underside (or face). The plane iron is that part of the cutting edge or knife. A section in the front provides an escape for the shavings.

There are various types of planes (such as jack, jointer, smoothing, block, molding, rabbet, grooving, and router). The jack plane is for very heavy, rough work. It can be used as a smoothing or jointer plane that is a finer cut, if used properly.

Block planes are used primarily as a one-hand plane and almost exclusively for planing across the grain. The angle of the plane iron is different from other planes, and this is why it is generally called a low-angle block plane.

The Surform tool, which has a blade that looks like a cheese grater, is another useful tool in the plane family.

Files and rasps are tools used for smoothing surfaces such as metal and wood. Files are also used to sharpen saws and other cutting tools. Various shapes of files and rasps are used in carpentry work (such as flat, taper, square, round, and half-round).

Sandpaper is a tough paper or other material with abrasive glued to one side. It is used for smoothing and final finishing of wood surfaces. Various sizes and grades are on the market, ranging from extra fine to coarse. Sandpaper generally is purchased in 9-inch × 11-inch sheets, but can also be obtained in rolls and discs for various sanding machines.

Review Questions

1. What is a spoke-shave?
2. Name the various types of planes.
3. What is a scraper?
4. What is a plane iron?
5. What is a low-angle block plane?
6. How is the Surform used?
7. List five of the nine useful planes.
8. Is there any difference between the jack plane and the fore plane?
9. What is the difference between the smooth and block planes?
10. How is sandpaper used to smooth wood surfaces?

Chapter 10

Awls, Augers, Bits, and Braces

There are several kinds of boring tools, each class adapted to meet special working conditions, such as the following:

- Boring
- Countersinking
- Drilling
- Enlarging
- External boring (turning)
- Punching

The various kinds of tools used for these operations are brad awls (boring), gimlets and augers (countersinking), drills (drilling), hollow augers and spoke pointers (enlarging), countersinks (external boring), and reamers (punching). These various tools are called *bits* when provided with a *shank* instead of a *handle* for use with a *brace*.

Awls

An awl is a pointed tool for piercing small holes. The blade is shaped with a point to suit the conditions of use.

Brad awls have an edge like a screw driver and can be used as such on small screws (see Figure 10-1). The brad awl's principal use is in quickly making a hole for starting a nail or screw into hard wood. Brad awls are not commonly manufactured today. Make your own by regrinding a small screwdriver. Form a flat sharp taper with two bevels. Hone the edge with a smooth-cut file or stone.

A scratch awl is a pointed tool that is used for small starting holes for screws and nails. It can also be used to accurately scribe a line (see Figure 10-2).

Augers

Augers are used for boring holes from $\frac{1}{4}$ inch to 2 inches. When made with a shank for use in a brace, this style of auger is commonly called a bit (see Figure 10-3).

Because of the enormous variety of bits on the market, it is difficult to select the best one for a given purpose. The bits look alike. For example, Figure 10-4 shows an enlarged view of just two common styles of wood bits. One has a screw point, the other has a brad, or diamond, point. Note also that one style has a solid center with

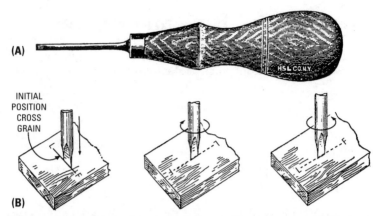

(A)

INITIAL
POSITION
CROSS
GRAIN

(B)

Figure 10-1 (A) Brad awl. Sizes $1/16$ to $3/16$ diameter. (B) Method of using the brad awl. Always start with the edge of the tool across the grain of the wood. In forcing the tool into the wood, do not turn the tool completely around, but give only sufficient turning movement in alternate directions to cut and crush the fibers, extreme positions LF, and the arrows indicating this movement.

Figure 10-2 The point of a scratch awl is conical.

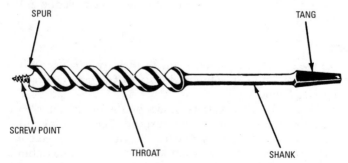

SPUR

TANG

SCREW POINT

THROAT

SHANK

Figure 10-3 A typical auger bit showing its component parts.

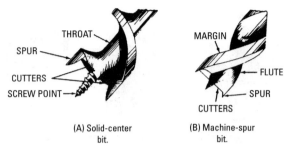

(A) Solid-center
bit.

(B) Machine-spur
bit.

Figure 10-4 Two common types of wood bits. The cutting edges of these bits are quite similar. The opening between the spiral is called the throat, but in some styles of double-twist bits, it is called the flute. Both terms are used alternately and mean the same thing.

a single spiral running around it, while the other is a double-spiral twist bit. However, they do look alike. The information given here will help, and so will experience (Figures 10-5 through 10-8).

If it is necessary to drill holes to an exact depth, an adjustable bit gage, such as the one shown in Figure 10-9, may be used. This tool is simply a clamp that can be securely attached to any standard-size wood bit by means of two wing nuts.

Among other types of bits frequently used in woodworking shops are router bits, end cutters (for cutting rosettes, rounding, and shaping), and a variety of other designs. Bits may also be used to cut other soft materials (such as Plexiglas).

To sharpen the spur of an auger, hold the bit in the left hand with the twist resting on the edge of the bench. Turn the bit around until the spur to be sharpened faces up. File the side of the spur next to the screw, carefully keeping the original bevel. File lightly until a fine burr shows on the outside, which is carefully removed by a slight stroke with a file. The result is a fine cutting edge.

To sharpen the cutter, hold the bit firmly in the left hand with the worn point down on the edge of the bench, slanted away from the hand with which you file. File from the inside back, and be careful to preserve the original bevel. Take off the burr or rough edge. Never sharpen the outside of the spur.

It is rarely necessary or advisable to sharpen the worm. However, it may often be improved if it is battered by using a three-cornered file that is carefully manipulated. Use a size that fits the thread. A half-round file is best for the lip and, with careful handling, may be used for the spur. Special auger-bit files are available for this purpose.

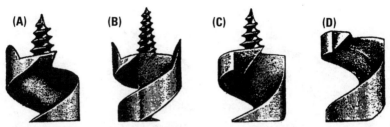

Figure 10-5 Bit and auger heads. (A) Single cutter, extension lip, coarse screw. Recommended for general all-around boring. Rapid, clean cutting, and very easy boring. Particularly adapted for difficult boring in wet, green, very hard, or knotty wood and boring with the grain. (B) Double cutter, extension lip, fine screw. Recommended for furniture and cabinet work, or wherever a particularly smooth hole is essential. Bores easily and clears readily. (C) Ship head with single cutter and coarse screw. Note absence of lip. Recommended for deep boring or in woods with strong grain. Especially adapted for boring plug holes in making riveted copper fastened joints in fine boat construction. Will stand many sharpenings. Does not bore as smooth a hole as types with spur. (D) Ship head single cutter without either screw or lip, sometimes called barefoot. Especially recommended for deep boring in wet pitchy woods, or when particularly straight boring is essential. Having no screw, it has little tendency to follow or drift with the grain of the wood.

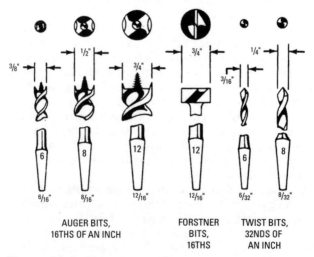

Figure 10-6 Typical auger, Forstner, and twist bits and the methods of marking their size.

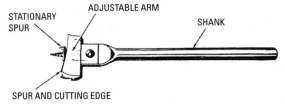

Figure 10-7 A typical expansive auger bit with an adjusting screw. The expansive bit obsoletes the necessity for many large bits. The cutter may be adjusted for various size holes. The size of the hole to be cut may be reduced or enlarged $1/8$ inch by turning the adjusting screw one complete revolution in the direction desired. Test for the correct size setting on a piece of waste wood before boring the hole in the wood being worked.

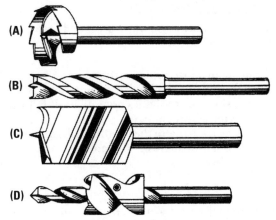

Figure 10-8 Special wood-bit styles. The bits represented are a multispur bit (A), a double-spur bit (B), a center bit (C), and a countersink bit (D). Bits not equipped with feed-screw points are usually meant to be used in the drill press, whereas bits equipped with screw points are for use with the hand brace.

Figure 10-9 A typical depth gage and its use. An adjustable bit gage of this type may be used to regulate the depth of the holes to be bored.

Twist Bits

In addition to augers and gimlets, a carpenter should have a set of *twist bits*. These tools are used for drilling small holes where the ordinary auger or gimlet would probably split the wood. They come either with square shanks for use with bit braces, or with straight shanks for use with electric drills (see Figure 10-10). These drills are available in standard sizes from $\frac{1}{16}$ to $\frac{5}{8}$ inch or more, varying in size by $\frac{1}{32}$ of an inch.

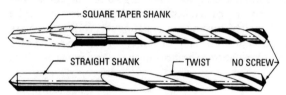

Figure 10-10 A bit stock twist drill for use with a brace and a straight-shank twist drill for use in the drill press.

A twist bit differs from an auger or gimlet in that it has no screw and has a less acute cutting angle of the lip. Therefore, there is no tendency to split the wood, since the tool does not pull itself in by a taper screw but enters by external pressure.

For many operations, especially where the smaller drills are used (as in drilling nail holes through boat ribs and planking), a geared breast drill is preferable to a brace.

Countersinks

Sometimes it is necessary to make a conical enlargement of a hole at the surface of the wood. This operation is performed by a bit tool called a *countersink*. The countersink may be used in a hand brace, an electric drill, or a drill press (see Figure 10-11).

Figure 10-11 A typical rose countersink.

Hand Drills and Braces

The hand drill and brace (see Figure 10-12) are the conventional hand tools used by the carpenter for holding and turning bits. The hand drill is used for rapid drilling of small holes. The brace differs from the hand drill mainly in that, with the brace, the turning movement is applied directly to the bit by means of the handle swing. The hand drill is equipped with a gear-pinion arrangement for turning the drill (see Figures 10-13 and 10-14).

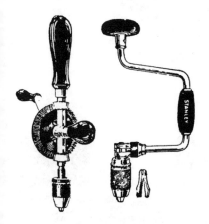

Figure 10-12   A hand drill and a hand brace.

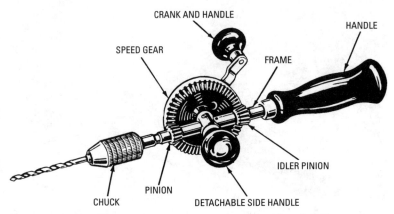

Figure 10-13   The component parts of a typical hand drill.

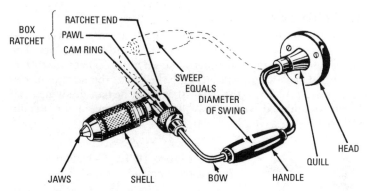

BOX RATCHET
RATCHET END
PAWL
CAM RING
SWEEP EQUALS DIAMETER OF SWING
HEAD
QUILL
JAWS
SHELL
BOW
HANDLE

Figure 10-14 Parts of a typical hand brace.

Figure 10-15 The proper method of boring a vertical hole with the hand brace. The bit must be perpendicular to the work surface.

(Courtesy of Stanley)

Using Boring Tools

Satisfactory results in the use of boring tools are only obtained with practice and the use of good tools, each suitable for the particular job assigned to it. The work should be properly laid out, and the hole should be clearly marked. To bore a vertical hole, hold the brace and bit perpendicular to the surface of the work (see Figure 10-15). Compare the direction of the bit to the closest straight edge or to the sides of the vise. A try square may also be held near the bit to be certain of the true vertical position.

To bore a horizontal hole, hold the head of the brace cupped in the left hand against the stomach, with the thumb and forefinger around the quill (see Figure 10-16). To bore through the wood without splintering the second side, stop when the screw point reaches the other side, and finish the hole from that side. When boring with an expansive bit, it is best to clamp a piece of scrap wood to the second side and bore straight through.

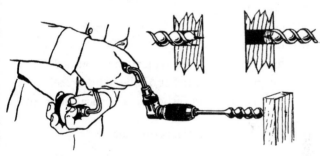

Figure 10-16 The correct method of boring holes horizontally to prevent splitting and splintering.

Frequently, restricted working quarters make it necessary to use the ratchet device of the hand brace. The ratchet brace is indispensable when boring a hole in a corner, or when a projecting object prevents the user from making a full turn with the handle. To actuate the ratchet, turn the cam ring. Turning the cam ring to the right will allow the bit to turn clockwise and give a ratchet action when the handle is turned left. Turning the cam ring to the left will reverse this action.

Summary

Various kinds of boring tools are used in woodworking shops. Tools such as punches, drills, countersinks, enlargers, and boring implements are used every day in various operations. These tools are generally called bits when provided with a shank instead of a handle for use with a brace or for use in an electric drill.

There are various types and sizes of augers. Some augers have a screw point and others have a brad or diamond point.

Twist bits are used for drilling small holes. They are designed with square shanks for brace or straight shanks for electric drills. The twist bit differs from an auger in that it has no screw and has a less acute cutting angle of the lip. Therefore, there is less danger of splitting the wood.

Review Questions

1. Explain the auger and its uses.
2. What is the difference between the flute and throat on a wood bit or auger?
3. What is the spur of a wood bit?
4. List four kinds of boring tools.
5. Describe the brad awl.
6. For what is the scratch awl used?
7. What does the screw point on a typical auger bit do?
8. What is the purpose of the ratchet on the hand brace?
9. Describe the proper method of boring a vertical hole with a hand brace.
10. What are the three types of points on augers?

Chapter 11

Hammers, Screwdrivers, Wrenches, and Staplers

Fastening tools are hammers, wrenches, screwdrivers, and the like. They are used for driving the following:

- Screws
- Nails
- Brads
- Staples

Hammers

The *hammer* is an important tool in carpentry, and there are numerous types to meet the varied conditions of use. All hammers worthy of the name are made of the best steel available and are carefully forged, hardened, and tempered.

The shapes of the claws of hammers vary slightly in the products of different manufacturers, though they may all be called *curved-claw hammers*. Carpenters often develop a preference. The *straight-claw* (or *ripping*) *hammer* is not as popular as the curved-claw hammer (see Figure 11-1) with most workers because it does not grip nails for withdrawal quite so readily, and thus nails cannot be withdrawn easily. The shape of the poll is immaterial, but the octagon-shaped poll seems to be most popular. All good hammers have slightly rounded faces, thereby making it possible to drive a nail head down flush with the wood without unduly marking the material.

Several types of hammers are currently available (see Figures 11-2 through 11-5). One popular type is forged (head and handle) of a single piece of steel with the grip built up of leather or neoprene. Fiberglass-handle hammers have become popular also. The handle is made of polyester resin reinforced by continuous fiberglass in parallel form giving excellent strength. These hammers are available in 13-, 16-, and 20-ounce nail types, as well as 16- and 20-ounce ripping types, with nonslip neoprene grips.

Although metal and fiberglass hammers are popular, many workers still prefer the old standard adze-eye hammer with a handle of springy hickory. True, such handles can become sprung, or perhaps break, but the feel of such hammers is something some craftsmen do not want to part with. Without a doubt, anyone can become accustomed to the metal-handled tools in time.

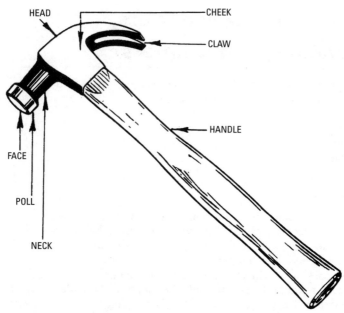

Figure 11-1 A typical bell-faced nail hammer.

Figure 11-2 Full polished octagon neck, round face, air-cushioned neoprene grip hammer. This hammer is forged steel from head to toe with a hickory plug in the head to absorb shock.

(Courtesy Vaughan and Bushnell Mfg. Co.)

The beginner should select a good hammer from an established and recognized manufacturer. The advice of an experienced carpenter is also extremely desirable. At any rate, the novice will bend a sufficient number of nails with the best of hammers until acquiring the skill in using this most indispensable tool. It is not a good policy for the neophyte to change to a lighter hammer when doing trim work. The beginner will usually be slow enough to acquire good control of one hammer. Builders and other workers who do a large amount

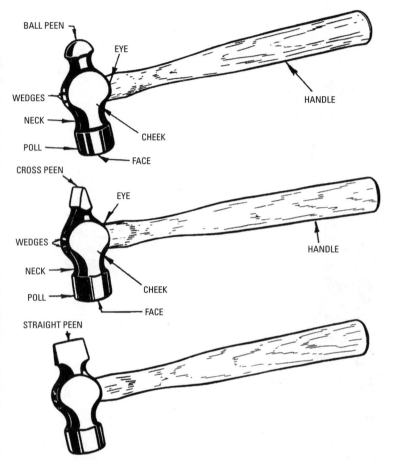

Figure 11-3   Ball-, cross-, and straight-peen hammers.

of heavy spiking often use a 2-pound hammer, sometimes with a cut face. However, such hammers are of no use for anything else.

When using a hammer, the handle should be grasped a short distance from the end, and a few sharp blows rather than many light ones should be used when driving nails. Keep the hand and wrist level with the nail head so that the hammer will hit the nail squarely on the head instead of at an angle. Failure to do this is the reason for difficulty so often experienced in driving nails straight.

The hammer face must be clean and dry to drive a nail straight (see Figures 11-6 and 11-7). Therefore, rub the face of the hammer

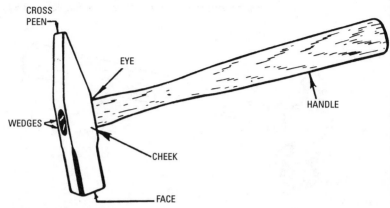

Figure 11-4 A riveting hammer.

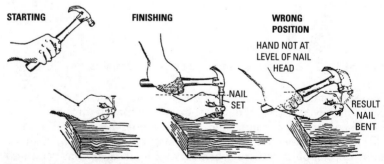

Figure 11-5 A typical soft-faced hammer.

Figure 11-6 Method of driving a nail. Guiding nail with left hand at the start. Using nail set to drive nail head below surface of the wood and to prevent hammer marks. The last illustration shows the result of not hitting nail square because of the wrong position of the hand. The dotted line shows nail knocked sidewise.

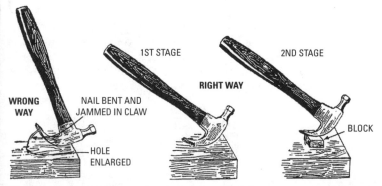

Figure 11-7   Wrong and right methods of drawing a nail with claw of the hammer.

frequently on wood. Hammers are designed to drive nails and not to hit wood (or fingers). Thus, when starting, tap gently while the nail is guided with the fingers and finish with a nail set. Hammers vary in size from 5 to 20 ounces. Most carpenters use a hammer weighing 14 to 16 ounces.

It is always a good idea to protect eyes and face from dust and flying particles when using a hammer. Many safety goggles are to be worn over personal glasses, with large clear-vision lenses and air holes for maximum ventilation (see Figure 11-8).

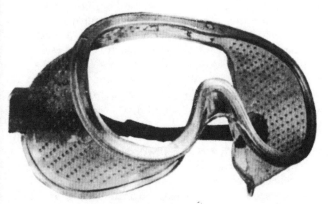

Figure 11-8   Safety goggles for use with striking tools.
(Courtesy Vaughan and Bushnell Mfg. Co.)

Screwdrivers

A screwdriver is similar to a chisel and usually differs mainly in the working end. The end is blunt on a screwdriver. Few screwdrivers have an ideally shaped end. Usually the sides that enter the slot in the screw are tapered. This is done so that the end of the screwdriver will fit into screw slots of widely varying sizes (see Figure 11-9).

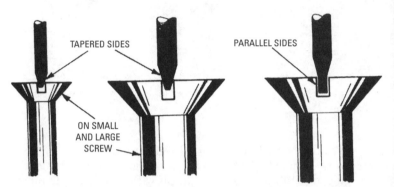

Figure 11-9 The end of a screwdriver should be shaped so that its sides are parallel. A screwdriver whose end is tapered can be used, but considerable downward pressure must be exerted to prevent the screwdriver from rising out of the screw slot. With parallel sides, there is no tendency for the screwdriver to rise, no matter how much turning force is exerted.

When using a screwdriver with a tapered blade tip, a force is set up because of the taper. That force tends to push the end of the tool out of the slot. Therefore, it is better to have several sizes with properly shaped parallel sides than to depend on one size with tapered sides for all sizes of screws (see Figure 11-10).

The operation of driving a screw with a plain screwdriver consists of giving it a series of half turns. Where a number of screws are to be tightened, time can be saved by using a screwdriver bit that is used with a brace in the same manner as an auger bit.

The quickest method of driving a screw using hand tools is by means of the automatic ratchet-type screwdriver (see Figure 11-11). The advantage of this type over the plain screwdriver is that instead of grasping and releasing the handle from 25 to 30 times in turning a screw home, it is grasped once and with two or three back-and-forth strokes, the screw is driven home, thus saving labor and time. The screwdriver drives or withdraws screws according to the position of

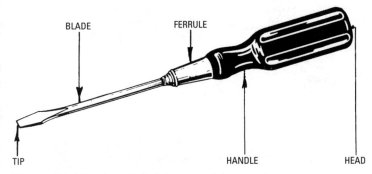

Figure 11-10 A typical plain screwdriver.

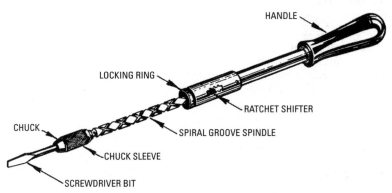

Figure 11-11 A spiral-ratchet screwdriver.

the ratchet shifter because pressure on the handle causes the spindle and tip to rotate. The ratchet shifter can also lock the screwdriver into a rigid unit, so it can be used like a conventional screwdriver.

Special ratchet-type screwdrivers may be obtained with spirals of different angles to suit working conditions (such as a 40° spiral for rapidly driving small screws, a 30° spiral for general work, and a 20° spiral for driving large screws in hardwood).

The Phillips head screwdriver (see Figure 11-12) is constructed with a specially shaped blade tip to fit Phillips cross-slot screws. The heads of these screws have two perpendicular slots that intersect each other in the center. This design checks the tendency of some screwdrivers to slide out of the slot and on to the finished surface of the work. The Phillips screwdriver will not slip and burr the end of the screw if the proper size tool is used.

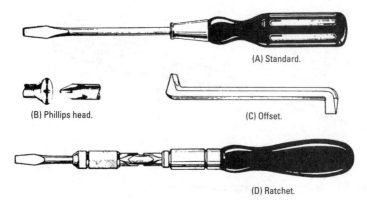

(A) Standard.

(B) Phillips head.

(C) Offset.

(D) Ratchet.

Figure 11-12 A typical assortment of screwdrivers.

The offset screwdriver in Figure 11-12 is a handy tool to use in tight corners where working room is limited. It usually has one blade forged in line with the shank, or handle, but this is not always the case. The other blade is at a right angle to the shank. The ends of the screwdriver can be changed with each turn, thereby working the screw into or out of its hole. This type of screwdriver is normally used only when the screw location is such that is prohibits the use of a plain or spiral-ratchet screwdriver. The offset screwdriver is also available in the ratchet form.

Wrenches

There are three general classes of wrenches:

- Plain
- Adjustable
- Socket

Plain (or *open-end*) *wrenches* are of the solid, nonadjustable type. They have fixed openings at each end (see Figure 11-13). A conventional set of wrenches, designed for the carpenter, use between 8 and 12 wrenches ranging in size from $5/16$ to 1 inch in jaw width.

Another form of plain wrench is the *box-end wrench* (see Figure 11-14). These are so called because they "box," or surround, the nut or bolt head. Their usefulness is based to some extent on their ability to operate in close quarters. Because of their unique design, there is little chance of the wrench slipping off the nut. Twelve notches are arranged in a circle around the inside of the box. This

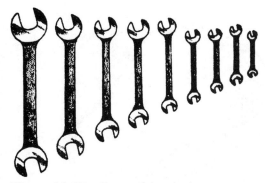

Figure 11-13 Open-end wrenches.

Figure 11-14 A typical box-end wrench.

12-point wrench can be used to continuously loosen or tighten a nut with a handle rotation of only 15°, as compared to the 60° swing required by the open-end wrench if it is reversed after each swing. A combination wrench employs a box-end wrench at one end and an open-end wrench at the other. Both ends are usually the same size, although this is not always the case.

Adjustable wrenches are somewhat similar to open-end wrenches. The primary difference is the one jaw is adjustable (see Figure 11-15). The angle of the opening with respect to the handle of an adjustable wrench is 22.5°. A spiral worm adjustment in the handle permits the width of the jaws to be varied from zero to 1 inch or more, depending on the size of the wrench. Always place the wrench on the nut so that the pulling force is applied against the stationary jaw. Tighten the adjusting screw so that the jaws fit the nut snugly.

The wrench most commonly found in a carpenter's toolbox is probably the old-fashioned *monkey wrench* (see Figure 11-16). For most uses, this wrench fills the bill well. Fit the jaws of the wrench snugly on the nut to be turned. It is usually advisable to turn *toward* the screw side of the handle because the wrench is then not so likely to slip off the nut. Do not try to turn the nut or bolt with the tips of the jaws or slip a piece of pipe over the handle to increase leverage.

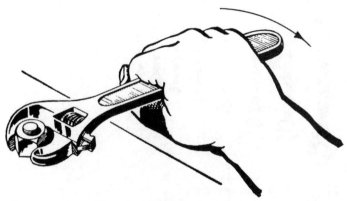

Figure 11-15 The adjustable wrench and the method of tightening a nut.

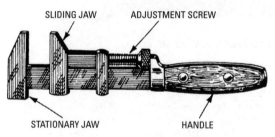

Figure 11-16 A typical monkey wrench.

A Stillson pipe wrench may be able to take it, but an ordinary monkey wrench won't. Keep the adjustment screw well oiled.

The wrench can be a somewhat dangerous tool, especially when great force is applied to move an obstinate nut. Under such conditions, the jaws often slip off the nut, thereby resulting in injury to the worker by violent contact with some metal part.

Other Fastening Tools

There are a number of other fastening tools to know about. The Pop rivet gun is good for securing thin materials, such as sheet metal, where only one side is accessible. You just put a rivet in the head, stick the rivet in a hole that passes through both parts, and squeeze—the rivet clamps them together.

The hand *staple gun* (an electric model is also available) is great for rapid application of staples when fastening ceiling tile and the like (see Figure 11-17). Another stapler that is operated by being rapped against a surface, similar to a hammer, is also available. It is popular with insulation installers.

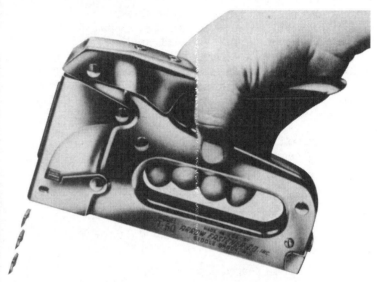

Figure 11-17   A staple gun. *(Courtesy of Arrow Fastener Co.)*

A close cousin of the stapler is the *nailing gun*, a recent addition to the hand-tool field that looks like a stapler and enables the worker to drive and countersink brads at the same time for easier installation of paneling.

Summary

Fastening tools include various hammers, screwdrivers, and wrenches. These tools are used for securing or joining various parts of wood or other materials together with nails, screws, or bolts.

The hammer is a very simple striking tool. It is made in a number of sizes and shapes to perform specific tasks. All hammers worthy of the name are made of the best steel, carefully forged, hardened, and tempered. There are various groups or classifications, such as nail or claw, ball peen, soft face, cross and straight peen, tinners, and riveting. The 16-ounce size is most popular for driving nails into wood.

Screwdrivers are designed to insert or remove screws from various materials. There are several classes of screwdrivers used in the average shop (such as plain, Phillips, offset, and ratchet). The plain screwdriver is very similar to a chisel except it has a blunt end. The Phillips screwdriver is made with a specially shaped blade tip that fits Phillips cross-slot screws.

The wrench is a tool for tightening or loosening bolts and nuts used in the assembly of numerous articles of wood or other material. The majority of nut and bolt heads are hexagonal (six-sided), although other shapes are sometimes encountered. The wrench is designed to grip nuts and bolt heads and then turn them by means of lever action exerted at the handle. Various types of wrenches manufactured are open end, monkey, adjustable, box end, and combination. Other fastening tools include the rivet gun and staple gun.

Review Questions

 1. What is a Phillips screwdriver?
 2. How are hammers sized for a particular job?
 3. Name the various types of wrenches manufactured.
 4. Name the various types of hammers manufactured.
 5. What are the requirements for a good screwdriver?
 6. For what are staples used?
 7. Where is the soft-faced hammer used?
 8. Why is using the right screwdriver for the screw head important?
 9. Why should safety goggles be used when driving nails?
10. What is the difference between a box end and an open-end wrench?

Chapter 12

Portable Power Tools

Portable power tools have become the mainstay of the modern carpenter. They can increase productivity on the job site, but they do not guarantee it. Improperly handled, power tools can cause much worse accidents than muscle-powered tools. Careful selection of the correct tool for the job and training in its proper use ensures safe and accurate work.

Cordless Power Tools

Cordless, battery-powered tools give the busy carpenter freedom to move quickly about the construction site without having to fiddle with extension cords. Ten or 15 years ago, when cordless tools first came out, they were more of a nuisance with little practical advantage over corded portable tools. Batteries needed careful nursing and frequent charging. Now, battery and electronics technology have caught up with practical needs of the carpenter on the job site (see Figure 12-1).

The new 18-volt tools provide important advantages over earlier 7.2-volt, 9.6-volt, and 14.4-volt systems. Longer battery life means work continues with less frequent stops for battery changing and charging. The 30-minute to 1-hour charging times means batteries can be charged during a coffee or lunch break. Earlier systems needed at least a few hours or overnight charging.

There are three main components in a portable power tool system: the tool itself, the battery, and the charger. A cordless drill has a plastic housing that contains a battery holder, switch, motor, gearbox, and chuck (see Figure 12-2). The battery is a set of storage cells in a plastic housing. The battery housing is shaped so it will fit into the tool and charger in only one way, to make certain the positive and negative terminals are correctly connected. The charger has electronic circuits that feed electricity from a wall outlet into the battery at the correct level and for the right amount of time. When the battery is full, the circuits cut back the level of charge to a trickle that maintains a full charge until the battery is removed. In the most recent models, the battery pack is attached to the bottom of the handle. The large battery bulge is placed there so the drill will be balanced when you use it. It does become somewhat awkward when using to drill a vertical hole.

Generally speaking, the higher the battery voltage, the more power available for starting torque. That means the drill will be able to bore through tougher wood and/or metal.

Figure 12-1 Cordless drill. This 12-volt, high-torque drill has a variable, reversing, two-speed drive with a $^3/_8$-inch chuck. The bulk and mass of the battery at the end of the grip balances the weight of the motor and gears in the main housing. *(Courtesy Milwaukee Electric Tool)*

Advantages

The main advantage of using cordless tools is the freedom of movement they allow. To install a gutter you could go up a ladder, screw brackets down walking along the edge of the roof. Then you come down a ladder on the other side of the house and are ready for the next task. All this can be accomplished without running extension cords that would be needed with ordinary power tools. Cordless tools are especially suitable for repeated tasks requiring little power (such as installing wallboard screws or spot sanding filled nail holes). You might get a few hundred to several hundred operations on a single charge.

Disadvantages

Cordless tools have limited power output. Drilling 1-inch diameter holes in oak timbers would be too much work for a cordless

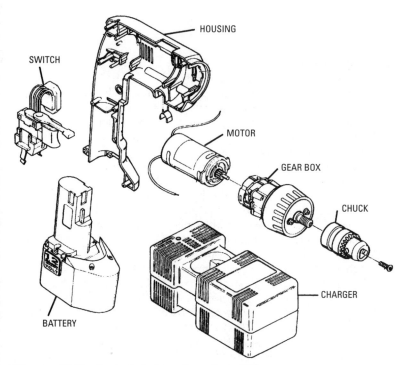

SWITCH

HOUSING

MOTOR

GEAR BOX

CHUCK

CHARGER

BATTERY

Figure 12-2 Exploded view of cordless drill. *(Courtesy Milwaukee Electric Tool)*

drill. It might only drill two or three holes before the batteries were drained. The higher voltage drills do overcome some of this with higher starting torque.

Cordless tools are usually more expensive to buy than corded tools. However, if you can use them efficiently, you will recover the added expense in saved time.

Standard Electric Tools

Portable power tools that plug directly into electric outlets provide considerably more power for work. While some manufacturers provide tools with grounded plugs, others make them with plastic housings. Both methods provide a defense against electric shock. Do not buy or use tools that do not use one of these methods for shock protection.

Before connecting the cord plug to an electrical outlet, be sure that the line voltage is the same as that stamped on the nameplate of the tool. When an extension cord is necessary, it should be of sufficiently heavy wire to ensure full voltage at the tool with the machine under load.

Electric Drills

The process of drilling holes in metal and boring holes in wood with an *electric drill* is similar to drilling or boring by hand, except that the power for turning is furnished by an electric motor instead of by the operator. Drills of this type usually have capacities for drilling holes from $1/16$ inch up to 3 inches in diameter.

To use an electric drill, first mark the location of the hole. Then, with the motor running, set the point of the drill on the mark and start drilling. Care must be used to hold the electric drill at right angles to the work so that the hole will be perpendicular to the surface. With the tool held in this manner, exert a light pressure and continue drilling. Use the variable speed feature as needed.

If the hole is to go completely through the work, relieve the pressure on the drill when the point of the drill bit begins to break through and until the hole is completed. Finally, withdraw the drill from the hole by pulling it straight back and then shut off the motor.

Twist drills do not pull themselves into the work. They must be fed by pressure, and this pressure must be exerted by the operator of an electric drill in exactly the same way as if drilling entirely by hand. The only effort saved the operator by the electric drill is that of turning. Electric drills are frequently used as a power source for a variety of attachments (see Figures 12-3 and 12-4).

Portable Saws

For the cutting of large panels, the *circular saw* (see Figure 12-5) is usually the way to go. These saws come in a variety of sizes with the size designated by the diameter of the blade the saw can accept. A blade about 7 inches in diameter is adequate for most carpenters. This means it can cut stock less than half the diameter of the blade (less than 3 inches).

Using the saw is not difficult, but as with all power tools, you have to be careful.

With the switch in the off position, rest the front edge of the saw base flat on the work. Start the saw by pulling back on the trigger switch and begin the cut, being careful not to jam the saw blade into the work suddenly. Forcing may place an unnecessary strain on the operating parts, possibly leading to their premature replacement. For most cutting operations, guiding the saw through the work is all that is necessary. If the motor should stall because of a dull blade or unnecessary pressure, do not release the switch at once, but pull

Figure 12-3 Electric drill fitted with an abrasive disk cutter.

the saw back, allowing the blade to run free before shutting off the motor. This precaution reduces burning of the contact points in the switch and greatly lengthens its life.

The *reciprocating* (or *back and forth*) *saw* shown in Figure 12-6 is a very handy tool to have on any construction job. The ability of this blade to cut through 2-inch lumber and follow a circle make it ideal for many curved cuts. It does, however, require two hands for safe operation.

Electric Planes

In finish carpentry work, the use of a plane is often necessary. For example, in fitting doors, it is usually necessary to plane one or more edges to obtain the proper clearance between the door and its jamb. A plane is also a necessity when cabinets are built on the job and for the installation of such items as bookshelves.

Electric planes are now available to perform much of the work formerly done with hand tools. An electric plane, such as the one

Figure 12-4 Electric drill fitted with a sanding disk.

shown in Figure 12-7, provides an accurate and rapid means of planing. In most instances, it will result in precision work with less effort. The model shown is adjustable for depth of cut and has a side fence that can be set to plane any desired angle. When set at 90°, the planed edge will be at a true right angle to the side of the material. The plane in Figure 12-7 has a 3-inch cutter that is more than adequate for most work.

Saber Saws
Another good addition to the carpenter's tool collection is the saber saw (see Figure 12-8). Most models are capable of cutting through

Figure 12-5 Circular saw with carbide-tipped blade.

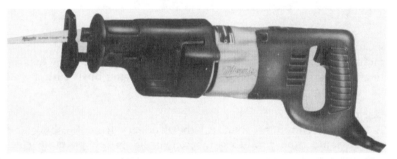

Figure 12-6 Reciprocating saw with two variable speed ranges. The small foot of this saw allows access to smaller spaces than a circular saw.
(Courtesy Milwaukee Electric Tool)

2-inch stock with ease (though it is better to use a circular saw for this) and can be used in places impossible to reach with an electric handsaw. The real value of the saber saw lies in the intricate shapes it can cut in plastic, wood, and even metal by using the proper type of blade. The cut can be adjusted to any desired depth and angle.

A full range of blades is available for this type of saw, from the standard combination blade for rough cuts in wood to metal cutting blades. Even knife-edged blades for cutting leather, fabric, and

Figure 12-7 Electric plane.

so forth may be purchased. Accessories are also available. These include fences for accurate straight-line cuts, circle-cutting attachments, and offset blade chucks to permit sawing flush with a wall or sawing up to an object.

Sanders

A tool that eliminates much of the drudgery from finishing and at the same time produces a better finish is the portable electric sander. There are many types of these handy machines available—the *orbital sander, reciprocating sander,* and the *belt sander.* Of the three, the belt sander (see Figure 12-9) provides the most rapid removal of material. Orbital and finishing sanders (see Figures 12-10 and 12-11) are better suited for final finishing work, however. All grades of grit are available for the belts and sheets used with these machines. The sander shown has a belt that is $4^{1}/_{2}$ inches wide.

Routers

A router consists of a canister-shaped housing with a bit projecting out of the base. The bit revolves at tremendous speeds. The operator grips handles on the sides of the canister and runs the router along the material (see Figure 12-12). The bit cuts.

Figure 12-8 A saber saw is used to make a blind cut in the middle of a plywood sheet.

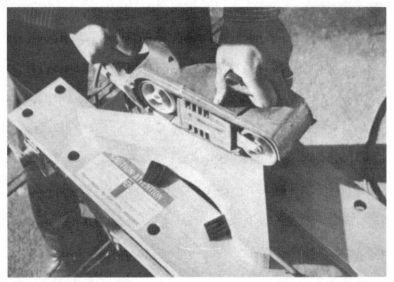

Figure 12-9 Belt sander.

Figure 12-10 The random orbit sander can smooth flat surfaces, as well as the contours shown here. The random action of the head helps eliminate swirl marks in the wood surface. *(Courtesy Porter-Cable)*

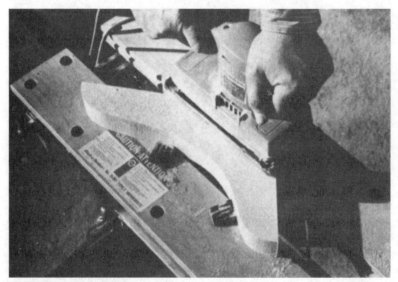

Figure 12-11 Finishing sander.

Routers can be fitted with a variety of bits to trim plastic laminate and make ploughs or grooves in boards and plywood.

Router bits come in a variety of materials. Carbide-tipped ones are more expensive but last far longer than plain steel and are acknowledged as well worth the extra cost when cutting plywood or particleboard.

Figure 12-12   Router.

Summary

Power-operated hand tools include electric handsaws, hand planes, routers and mortisers, air and electric hammers, portable drills, bench grinders, and hand sanders. Many of these tools have taken the disagreeable jobs out of carpentry work.

Power handsaws are very easy to operate and are made to cut material up to approximately 4 inches thick, depending on the size of the saw. It should be noted that, although very easy to operate, this type of power tool is potentially dangerous. Sudden jams and kickbacks can injure the operator. Generally, saws are equipped with combination blades for general-purpose work, but special blades for ripping, crosscutting, mitering, dadoing, metal cutting, and so forth, are available.

Drilling holes in metal or boring holes in wood with an electric drill is similar to drilling or boring by hand, except that power for turning is furnished by an electric motor instead of by the operator. Many drills can be fitted with attachments for driving screws, rotating small grinders, drilling at right angles, and so forth. Drills can be mounted on a drill stand, which provides an easy control for feeding the drill into the work.

Review Questions

 1. Name a few power-operated hand tools.
 2. What are some advantages in using a saber saw?
 3. Why are saw blades manufactured for the power handsaw?
 4. Name the three types of portable hand sanders.
 5. What are some advantages in using an electric plane?
 6. How long does it take a power tool battery to fully charge?
 7. What is the main advantage of using a cordless tool?
 8. What is the main disadvantage of using a cordless tool?
 9. Where would you use a reciprocating saw?
 10. Where is the electric plane used to its greatest potential?

Chapter 13

Tool Sharpening

It cannot be said too often that edged tools must always be kept in perfect condition to do satisfactory work. This means that the cutting edge must meet the following criteria:

- Be keen
- Be free of nicks
- Have the proper bevel

Sharpening is done by subjecting the tool to friction against an abrasive. The process includes grinding and honing.

Grinding

First, the tool is placed on a grinding wheel to bring the bevel to the correct angle and to grind out any nicks that may exist in the cutting edge (see Figure 13-1). Although this takes out the nicks and irregularities that are visible to the eye, the edge is still rough, as seen under a microscope. This roughness is considerably reduced by honing on an oilstone, although it is impossible to make the edge perfectly smooth because of the granular structure of the material.

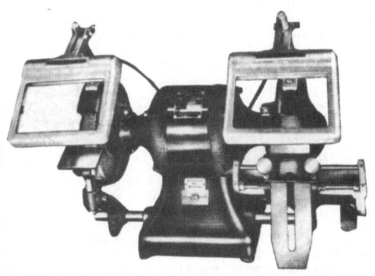

Figure 13-1   A bench grinder used to sharpen tools.

When grinding tools on a stone without the use of either a guide or a rest, the tool is firmly pressed to and held with both hands at an angle of approximately 60° on the face of the revolving stone (see Figures 13-2 and 13-3). Do not apply too much pressure (especially

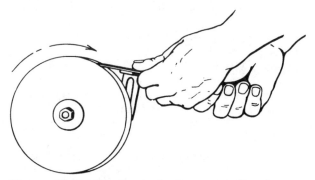

Figure 13-2 Grinding is the first step in the sharpening process. Different tools require different techniques. For a plane iron, the grinding wheel should turn toward iron.

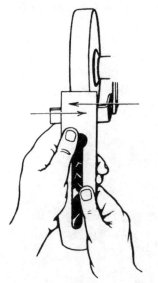

Figure 13-3 Moving plane from side to side grinds bevel down completely.

when grinding with the rapidly revolving emery wheels), since it is all too easy to burn the temper out of the tool. The edge of the tool must be continuously watched (especially with dry wheels). In case of a dry wheel, the tool must be immersed frequently in water to prevent overheating.

Plane-iron cutters vary in their make, temper, quality of steel, and uses, and they must be ground and sharpened for the sort of work they are intended to execute. As previously explained, it is usual to grind a jack plane iron slightly curved, and a jointer or smoothing plane iron flat, except at the corners (see Figure 9-13).

Before condemning any plane iron, therefore, carefully measure and compare the bevel of the cut and the thickness of the cutter. If the bevel is too long, the plane will jump and chatter.

If it is too short, it will not cut, so it must be ground to a proper bevel. The length of the bevel should be twice the thickness of the iron (see Figure 9-18).

Hatchets, axes, and adzes are always ground to their proper bevels. Some have double and others have single bevels (see Figure 8-15). When grinding, the blade is held to the stone surface with the right hand, and the handle is held with the left hand and on the left side, reversing the tool as the opposite side is being ground or sharpened.

Drawknives and spoke-shaves have cutters that should be held with both hands, and the blade should be kept horizontal on the stone as it is revolved toward the operator. Some woodworkers prefer to grind with the stone rotating toward the cutting edge, whereas others prefer to grind with the stone rotating away from the cutting edge. The latter is the safer method, because with the stone advancing, there is a danger of injury to the operator in case the tool digs into the stone. Do not use too much pressure with an advancing stone.

After the tool is ground, it should be honed.

Bench Grinders

Bench grinders are commonly used in woodworking shops for sharpening chisels, screwdrivers, and so on and for smoothing metal surfaces. Figure 13-1 shows a common type of bench grinder. This type of grinder consists mainly of an electric motor with a double-ended horizontal spindle, the ends of which are threaded and fitted with flanges to take the grinding wheels. Other models employ a conventional belt drive.

The size of the grinder is commonly taken from the diameter of the abrasive wheel used in connection with it. Thus, a grinder with a 6-inch diameter wheel is called a 6-inch grinder. Grinder units are further classified as bench or pedestal, the latter indicating a floor model.

A bench grinder is usually fitted with both a medium-grain and fine-grain abrasive wheel. The medium wheel is satisfactory for rough grinding where a considerable quantity of metal must be removed or where a smooth finish is not important. For sharpening tools or grinding to close limits of size, the fine wheel should be used because it removes metal slower, gives the work a smoother finish, and does not generate enough heat to anneal the cutting edges. When grinding tools, keep a pan of water handy and dip the tool in it often to ensure against overheating.

When a deep cut is to be taken on work, or when a considerable quantity of metal must be removed, it is often practical to grind with the medium wheel first and finish up with the fine wheel. The wheels are removable, and most bench grinders are made so that wire brushes, polishing wheels, or buffing wheels can be substituted for the grinding wheels.

Operation

In grinding, the work should be held firmly at the correct angle on the rests provided and fed into the wheel with enough pressure to remove the desired amount of metal without generating too much heat. The rests are removable, if necessary, for grinding odd-shaped or large work. As a rule, it is not advisable to grind work requiring heavy pressure on the side of the wheel because the pressure may crack the wheel. As abrasive wheels become worn, their surface speed decreases and this reduces their cutting efficiency. When a wheel becomes worn in this manner, it should be discarded and a new one installed on the grinder.

Safety Precautions

Before using a bench grinder, ensure that the wheels are firmly held on the spindles by the flange nuts and that the work rests are tight. Wear goggles, even if eye shields are attached to the grinder, and bear in mind that it is unsafe ever to use a grinder without wheel guards. Wearing glasses does not take the place of wearing goggles. A pair of expensive glasses is easily ruined, as the flying sparks stick to the glass and sometimes cannot be polished off. Also, remember that it is easy to run a thumb or finger into the wheel.

Oilstones

Oilstones are used after the grinding operation to give the tool the highly keen edge necessary to cut wood smoothly. The oilstone is so called because oil is used on it to carry off the heat resulting from friction between the stone and tool and to wash away the particles of stone and steel that are worn off by the rubbing. The process of rubbing the tool on the stone is called *honing*.

Honing

After a tool has been ground on a grinding wheel, it will still have a wire edge. This edge must be removed, and the edge must be further treated by honing on an oilstone (see Figures 13-4 and 13-5). The oilstone is constantly needed during all operations in carpentry in which the plane and chisel are used. It is needed more frequently than grinding because grinding is only necessary when the tool becomes

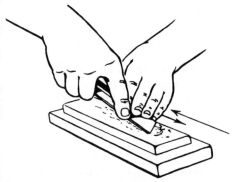

Figure 13-4 After grinding, oilstone is used. Plane iron is moved back and forth on stone with back edge slightly raised. Stone is kept moist with oil.

Figure 13-5 Wire or feather edge is removed with plane iron held flat on stone.

nicked, or when the edge becomes too dull to be sharpened on the oilstone without an undue amount of labor. The size of a typical oilstone for general use is approximately 2 inches × 8 inches, or 2 inches × 9 inches.

Using Oilstones
One rather desirable stone is the double carborundum (that is, a carborundum oilstone with one side coarse and the other side fine). Begin to hone on the coarse side and finish on the fine side with this type of stone. It is necessary to keep the oilstone clean and in perfect condition. The simple reality is that a stone that isn't in good

condition won't work properly. Oilstones should always be kept in their cases when not in use. Use only thin, clear oil on oilstones, and wipe the stone clean after using. Then moisten the stone with clean oil.

To clean an oilstone, wash it in kerosene. This will remove the gummed surface oil. This is easily done by heating the stone on a hot plate. A natural stone can also be heated on a hot plate to remove the surplus or gummed oil. After that, a good cleaning with ammonia will usually restore its cutting qualities. If this treatment does not work, scour the stone with a piece of loose emery or sandpaper. The sandpaper should be fastened to a perfectly smooth board or a piece of heavy plate glass. Clean the stone in a well-ventilated area and observe safety precautions.

When applying chisels and plane irons to an oilstone, hold the tool face up with both hands, the left hand in front, palm up, with the thumb on top and the fingers grasping the tool from underneath (see Figures 13-6 and 13-7). The right hand is held behind the left hand, palm down, with the thumb under and the fingers reaching across the face of the tool. The blade edge is then moved back and forth with a sliding rotary motion on the face of the stone (which is first lubricated with oil or water). The blade angle is generally approximately 60°. After 10 or 12 rubs, the blade is turned over and rubbed flat on the face side. The blade is then stropped; this may be done by rubbing on a piece of leather set on top of the oilstone case. When this is done, the keenness of the blade may be tested with the thumb or by drawing the edge across the thumbnail, but this test must be done carefully to avoid injury.

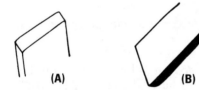

(A) (B) (C)

Figure 13-6 (A) When sharpening a plane iron, be aware that plane marks will show less on a surface if the iron's corners are slightly rounded. (B) When sharpening, do not rock the iron. A round bevel that doesn't cut well will occur. (C) A bevel on the flat side of the iron prevents the cap iron from fitting tightly, and shavings will clog the plane.

Outside gouges are sharpened in the same manner as chisels. The tool should be rolled on its axis when grinding the bevel. A

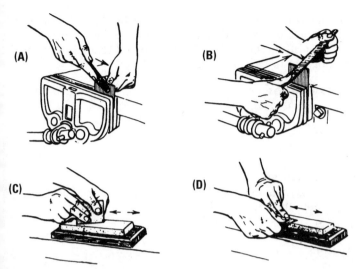

Figure 13-7 (A) To sharpen scraper blade on spoke-shave or any bevel-edge blade, first remove old burr by running a mill file against flat edge on flat side. (B) File or grind the bevel at about 45°. Push file forward and to the side. (C) Whet the bevel side of blade on oilstone. (D) Knock off burrs by rubbing face side against stone.

whetstone is used to remove the wire edge by rubbing on the inside concave surface. The curved edge of the whetstone must fit exactly the arc of each gouge as closely as possible. Inside gouges must be ground on a curved stone and whetted to keen edges with the oilstones and whetstones.

Hollows and rounds, beading, and other special plane cutters are usually sharpened with whetstones. They rarely require grinding. If they become nicked or damaged on their edges, they are utterly useless.

Cold chisels, punches, and nail sets are best sharpened or pointed on grinding wheels. Carving tools are sharpened with small, fine whetstones.

When honing or whetting fine bench chisels, the burnished-face side must be kept perfectly flat on the face of the oilstone by pressing firmly down with the fingers of the left hand. The handle is held in the right hand. The rubbing action must be gentle and rapidly repeated, turning the tool over constantly.

The edge of the chisel blade may slope slightly to the side of the oilstone, and it should be moved back and forth in a rotary motion on the stone. Do not raise the angle of the chisel too high on the stone, or the chisel will dig into and damage the surface of the oilstone. The oilstone should be wiped clean and reoiled frequently if several tools are to be sharpened.

Types of Oilstones
There are two general classes of oilstones:

* Natural
* Artificial

Natural Oil Stones
There are two general classes of natural stones grouped according to locality where they are found:

* Washita
* Arkansas

Washita Oil Stone
Washita stone is found in the Ozark Mountains of Arkansas and is composed of nearly pure silica. It is very similar to the Arkansas but much more porous. It is known throughout the world as the best natural stone for sharpening carpenters' and general woodworkers' tools. Its sharpening qualities are attributable to small, sharp-pointed grains or crystals, hexagonal in shape and much harder than steel. It is found in various grades, from perfectly crystallized and porous grit to vitreous flint and hard sandstone. The sharpness of grit depends entirely upon its crystallization. The best oilstones are made from very porous crystals.

Following are types of Washita oilstones:

* *Lily White Washita*—This is the best selection or grading of natural Washita. It is perfectly white in color, uniform in texture, and nicely finished.

* *Rosy Red Washita*—This has an even porous grit somewhat coarser than the Lily White grading and is therefore faster-cutting.

* *No. 1 Washita*—This is a good oilstone for general use, where a medium-priced stone is wanted. It is far superior to the many cheap so-called oilstones on the market (which are only sandstones with a polished face), but it is not as uniform as the Lily White.

Arkansas Oil Stone

Genuine Arkansas stone is composed of pure silica crystals, microscopic in size, and silica is among the hardest of known minerals. So hard and perfectly crystallized is the Arkansas stone that it is nearly 16 times harder to cut than marble. The hardest of steel tools with the finest points or blades may be sharpened on the Arkansas stone without grooving. Arkansas stone is prepared for commercial purposes in two grades:

- *Hard Arkansas*—This is much harder than steel and will, therefore, cut away and sharpen steel tools. The extreme fineness of texture makes it a slow cutter, but a perfect sharpener.

- *Soft Arkansas*—This is not as fine-grained and hard as the Hard Arkansas, but it cuts faster and is better for carvers, file makers, pattern makers, and all workers in hard wood.

Artificial Oil Stones

These are made of *carborundum*, emery, corundum, and other artificial abrasives. They are primarily used in place of natural stones because they cut faster and may be made of any degree of fineness and of even texture.

Carborundum Oil Stones

These are made from carborundum. They may be used dry, or with water or oil. They are quite porous, and may be tempered clean and bright. They never fill up or glaze over. They are made in three grades:

- *Fine*—Used for procuring a very smooth, keen edge on tools of hard steel and so forth.

- *Medium*—Used for sharpening tools quickly, where an extremely keen edge is not necessary.

- *Coarse*—Used for sharpening very dull and large tools that may later be finished with a fine stone or in cases where a fine finish is not required.

India Oil Stones

These are made from *alundum*. They possess the characteristics of hardness, sharpness, and toughness, as well as uniformity. They cut rapidly, and are especially adaptable to the quick sharpening of all kinds of machinists' tools and modern tool steels (such as scrapers, taps, reamers, milling cutters, lathe, and planer tools). All India stones are oil-filled by a patented process. This feature ensures a moist, oily sharpening surface with the use of only a small quantity of oil. It also ensures a good cutting surface by preventing the stone from filling with particles of steel.

India stones are made in three grades or grits:

- *Coarse*—This is for sharpening large and very dull or nicked tools, machine knives, and for general use where fast cutting is required without regard to fine finish.

- *Medium*—This is for ordinary sharpening of mechanics' tools not requiring a finished edge. It is especially recommended for tools used in working soft woods, cloth, leather, and rubber.

- *Fine*—This is for machinists and engravers, die workers, instrument workers, cabinetmakers, and all users of tools requiring a very fine, keen edge.

Summary

It is always important to keep a sharp edge on woodworking tools to do the best work. The cutting edge must always be free from nicks and have the proper bevel. Sharpening is done by grinding and honing.

The tools are placed on a grinding wheel to bring the bevel to the correct angle and to grind out the nicks and irregularities. After grinding, the tool is then honed on an oilstone to remove the roughness.

Review Questions

1. What angle is required on most cutting blades?
2. Explain the process of grinding the tools.
3. How do you clean an oilstone?
4. What is honing?
5. What happens to a tool that has been subjected to too much pressure when sharpening?
6. What is the first step in the sharpening process?
7. Hatchets, axes, and _____ are always ground to their proper bevels.
8. After a tool has been ground on a grinding wheel, it will still have a _____ edge.
9. An oilstone with one side coarse and the other side fine is known as _____.
10. Outside _____ are sharpened in the same manner as chisels.

Chapter 14

Saw Blade Sharpening

Saws can be sharpened professionally, but many carpenters like to do the job themselves. It's convenient, costs less, and a special kind of satisfaction is involved.

Besides the crosscut and ripsaw, a variety of other saws can be sharpened, including the backsaw, dovetail saw, keyhole saw, and coping saw.

There are five steps in the sharpening process:

- Jointing
- Shaping
- Setting
- Filing
- Dressing

When sharpening handsaws, the first step is to place the saw in a suitable clamp or saw vise (see Figure 14-1). The saw should be held

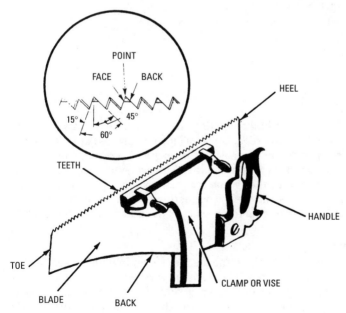

Figure 14-1   A method of fastening a handsaw in a saw clamp or vise.

tightly in the clamp so that there is no noticeable vibration. The saw is then ready to be jointed.

Jointing

Jointing is done when the teeth are uneven or incorrectly shaped, or when the teeth edges are not straight. If the teeth are irregular in size and shape, jointing must precede setting and filing. To joint a saw, place it in a clamp with the handle to the right. Lay a flat file lengthwise on the teeth, and pass it lightly back and forth over the length of the blade on top of the teeth until the file touches the top of every tooth. The teeth will then be of equal height, as shown in Figure 14-2. Hold the file flat. Do not allow it to tip to one side or the other. The jointing tool or handsaw jointer will aid in holding the file flat.

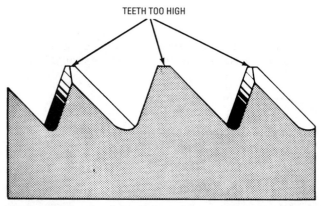

TEETH TOO HIGH

Figure 14-2 A method of jointing saw teeth. Place the saw in a clamp with the handle to the right. Lay a mill file lengthwise flat on the teeth. Pass it lightly back and forth along the length of the teeth until the file touches the top of every tooth. If the teeth are extremely uneven, joint the highest teeth first, then shape the teeth that have been jointed. Then joint the teeth a second time. The teeth will then be the same height. Do not allow the file to tip to one side or the other. Hold it horizontally.

Shaping

Shaping consists of making the teeth uniform in width, usually after the saw has been jointed. The teeth are filed with a triangular handsaw file to the correct uniform size and shape. The gullets (spaces between teeth) must be of equal depth. For the crosscut saw, the front of the tooth should be filed at an angle of 15° from the vertical, whereas the back slope should be at an angle of 45° from the vertical, as illustrated in Figure 14-3. When filing a ripsaw, the front of the teeth are filed at an angle of 8° with the vertical, and the back slope is filed at an angle of 52° with the vertical (see Figure 14-4). However, some good workers prefer to file ripsaws with more of an angle than this, often with the front side of the teeth almost square (or 90°). This produces a faster-cutting saw, but, of course, it pushes harder, and it will grab when cutting at an angle with the grain.

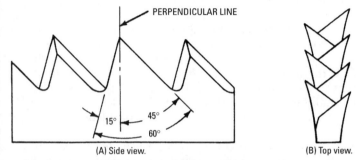

Figure 14-3 Side and tooth-edge views of a crosscut saw. The angle of a crosscut saw tooth is 60°, the same as that of a ripsaw. The angle on the front of the tooth is 15° from the perpendicular, whereas the back angle is 45°.

When shaping teeth, disregard the bevel of the teeth, and file straight across at right angles to the blade with the file well down in the gullet. If the teeth are of unequal size, press the file against the

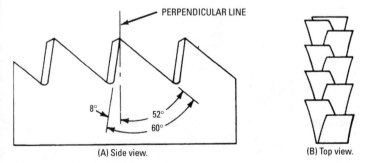

Figure 14-4 Side and tooth-edge views of a typical ripsaw. The tooth of a ripsaw has an angle of 60° (that is, 8° from the perpendicular on the front, and 52° on the back of the tooth).

teeth level with the largest flat tops until the center of the flat tops made by jointing is reached. Then move the file to the next gullet, and file until the rest of the flat top disappears and the tooth has been brought to a point. Do not bevel the teeth while shaping. The teeth shaped and of even height are ready to be set.

Setting

The teeth must be set after they are made even and of uniform width. *Setting* is a process by which the points of the teeth are bent outward by pressing with a tool known as a *saw set*. Setting is done only when the set is not sufficient for the saw to clear itself in the kerf. It is always necessary to set the saw after the teeth have been jointed and shaped. The teeth of a handsaw should be set before the final filing to avoid damage to the cutting edges. Whether the saw is fine or coarse, the depth of the set should not be more than one-half that of the teeth. If the set is made deeper than this, it is likely to spring, crimp, or crack the blade or break the teeth.

When setting teeth, particular care must be taken to see that the set is regular. It must be the same width along the entire length of the blade, as well as being the same width on both sides of the blade. The saw set should be placed on the saw so that the guides are positioned over the teeth with the anvil behind the tooth to be set (see Figure 14-5). The anvil should be correctly set in the frame, and the handles should be pressed together. This step causes the plunger to press the tooth against the anvil and bend it to the angle of the anvil bevel. Each tooth is set individually in this manner.

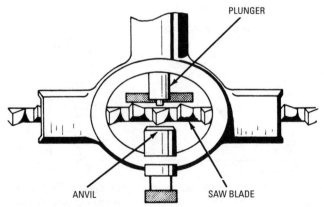

Figure 14-5 Position of the saw set on the saw for setting the teeth.

Filing

Filing a saw consists of simply sharpening the cutting edges. Place the saw in a filing clamp with the handle to the left. The bottom of the gullets should not be more than $1/2$ inch above the jaws of the clamp. If more of the blade projects, the file will chatter or screech. This dulls the file quickly. If the teeth of the saw have been shaped, pass a file over the teeth (as described in the section on jointing) to form a small flat top. This acts as a guide for the file. It also evens the teeth.

To file a crosscutting handsaw, stand at the first position (see Figure 14-6). Begin at the point of the saw with the first tooth that is set toward you. Place the file in the gullet to the left of this tooth, and hold the handle in the right hand with the thumb and three fingers on the handle and the forefinger on top of the file or handle. Hold the other end of the file with the left hand with the thumb on top and the forefinger underneath (see Figure 14-7).

Hold the file directly across the blade. Then swing the file left to the desired angle. The correct angle is approximately $65°$ (see Figure 14-8). Tilt the file so that the breast (the front side of the

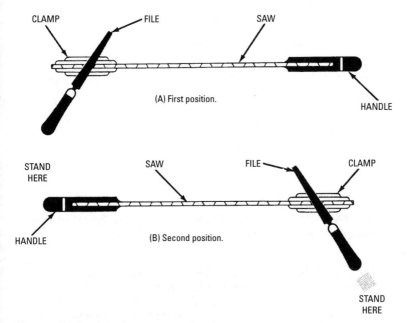

CLAMP FILE SAW

(A) First position.

HANDLE

STAND HERE SAW FILE CLAMP

HANDLE (B) Second position.

STAND HERE

Figure 14-6 Standing positions for filing a crosscut saw. The saw clamp should be moved along the blade as filing progresses.

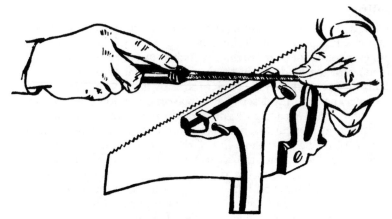

Figure 14-7 Method of holding the file when filing a handsaw.

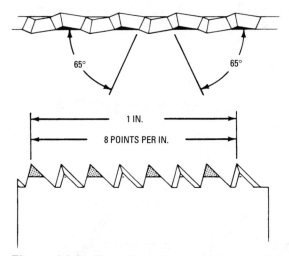

Figure 14-8 The side angle at which to hold the file when filing a crosscut saw having eight points per inch.

tooth) side of the tooth may be filed at an angle of approximately 15° with the vertical (see Figure 14-3). Keep the file level and at this angle. Do not allow it to tip upward or downward. Remember, files cut on the push stroke. They are raised out of the gullet on the reverse stroke. It cuts the teeth on the right and left on the forward stroke.

File the teeth until half of the flat top is removed. Then lift the file, skip the next gullet to the right, and place the file in the second gullet toward the handle. Now, if the flat top on one tooth is larger, then press the file harder against the larger tooth to cut that tooth faster. Repeat the filing operation on the two teeth that the file now touches, always being careful to keep the file on the same angle. Continue in this manner, placing the file on every second gullet until the handle end of the saw is reached.

Turn the saw around in the clamp with the handle to the left. Stand in the second position, and place the file to the right of the first tooth set toward you (see Figure 14-6). It is the first gullet that was skipped when filing from the other side. Turn the file handle to the right until the proper angle is obtained, and file the remaining half of the flat top on the tooth. The teeth that the file touches are now sharp. Continue the operation until the handle end of the saw is reached.

When filing a ripsaw, one change is made in the preceding operation. The teeth are filed straight across the saw at right angles to the blade. The file should be placed on the gullet to file the breast of the tooth at an angle of 8° with the vertical (see Figure 14-4). Stand in the positions shown in Figure 14-9. When sharpening a ripsaw, file every other tooth from one side. Then turn the saw around,

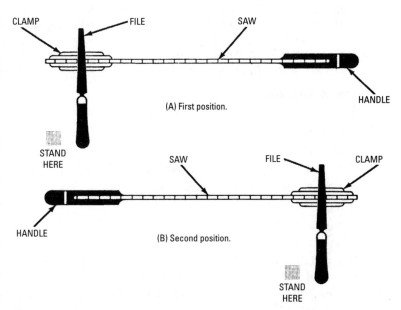

Figure 14-9   Standing positions for filing a typical ripsaw. Again, the saw clamp must be moved along the blade as filing progresses.

and sharpen the remaining teeth as described in the preceding paragraphs. When filing teeth, care must be taken in the final sharpening process to file all the teeth to the same size and height. Otherwise, the saw will not cut satisfactorily. Many good saw filers file ripsaws from only one side, taking care that the file is held perfectly horizontal. For the beginner, however, turning the saw is probably the most satisfactory method.

Dressing

Dressing of a saw is necessary only when there are burrs on the side of the teeth. These burrs cause the saw to work in a ragged fashion. They are removed by laying the saw on a flat surface and running an oilstone or flat file lightly over the side of the teeth.

Summary

The five steps that should be considered when sharpening saws are jointing, shaping, setting, filing, and dressing.

Jointing is done when the teeth on the blade are uneven or incorrectly shaped or when the teeth edges are not straight. The high spots or high teeth are filed down and then sharpened. Shaping a saw consists of making the teeth uniform. The teeth are filed to the correct size and shape. The gullets must be of equal depth.

After the teeth are made even and uniform in width, they must be set. This is accomplished with the use of a tool known as a saw set. It is necessary to set a saw when the teeth have been jointed and shaped. The saw must be placed in the clamp properly and held solid.

Review Questions

1. What are the five steps used in sharpening a saw?
2. What is the heel on a handsaw?
3. What is jointing a saw?
4. How are jointing and shaping accomplished?
5. Explain the process of dressing a saw.
6. What are the two types of handsaws?
7. Where is the gullet on a saw?
8. What is the kerf?
9. True or false: When filing a ripsaw, the teeth are filed straight across the saw at right angles to the blade.
10. Dressing of a saw is necessary only when there are ___ on the side of the teeth.

Chapter 15

Cabinetmaking

In making cabinets and other fine furniture, good joints that hold the parts together are essential. This chapter explores the tools, materials, techniques, and joints to accomplish this goal. Hand woodworking methods are emphasized. Learning to do this work by hand provides a deep understanding of the joint designs and of wood as a material. A worker with this fundamental knowledge will do better work when he or she moves on to using power equipment and more efficient production techniques.

Tools

A full set of good carpenter's tools is necessary in the cabinetmaking shop, including a set of firmer chisels, a set of iron bench planes, a set of auger bits with slow-feed screws that range in size from $1/4$ to 1 inch, and a set of numbered bits for the electric drill. In addition, the following tools will be found useful almost continually:

- Router plane
- Plow plane
- Miter box
- Several bar clamps (with varying lengths of bars)
- Hand clamps (those with wood jaws are most useful in the shop, but malleable C-clamps are often used)
- A $1/4$-inch electric drill

In addition to the regular vise, a workbench with an end vise for holding material between stops on top of the bench will be found to be quite convenient.

Joints

The most challenging job for anyone who works with wood is making joints. The problem is to cut two pieces of wood so that they meet without gaps—or with very little space—so the joint can do the job it is intended to do. This is a job requiring precision and patience. The ability to make good joints is the hallmark of a woodworker who has mastered the trade.

Before modern glues and clamps, many joints relied on wedges and keys. Stresses were also used to hold the joint together. Today, though, with tenacious glues and proper clamping procedures and equipment (plus the availability of special fasteners, as well as nails, screws, and bolts), the job can be done with relative ease. Indeed, most carpenters get by with just a few joints (such as butt, rabbet, dado, and miter).

Following is a presentation of the main joints used today, plus a consideration of some joints that the carpenter will rarely see but should know about as part of a well-rounded corpus of knowledge. For carpenters interested in working on antique furniture (or early buildings), knowledge of these joints can be important, even crucial, to do the job required.

There is a multiplicity of joints. They may be divided into general classes according to the manner in which the joining pieces are brought together.

Joints used in cabinetwork are usually classified according to their general characteristics as follows:

- Glued
- Beveled
- Hidden slot-screwed
- Coopered

Following are the categories of plain or butt joints:

- Straight
- Dowel pin
- Corner: square and miter
- Miter
- Feather
- Splice

Following are the categories of lap joints:

- Rabbet
- Dado (housed butt)
- Scarf
- Mortise and tenon
- Dovetail
- Tongue and groove
- Halved lap and bridle joints

Under each of these classifications are a number of joints that will be considered separately and will be briefly explained.

Glued Joints

In cabinetwork, practically all joints are (or should be) glued. Following are several types of glues suitable for use in the cabinet shop:

- *White Glue, (Polyvinyl)*—This is a most common woodworking glue. It is easy to apply, dries in about an hour, and is clear. Like most glues, it must be clamped until set.

- *Carpenter's Glue (Aliphatic)*—This works very much like white glue, but it dries much faster (in perhaps half the time). Also called *yellow glue,* it is not waterproof.

- *Contact Cement*—This is the adhesive favored for the application of plastic laminate. The adhesive is applied to both surfaces of things to be joined, allowed to get tacky, then pressed together. As the name suggests, the parts bond together instantly—indeed, there is really no room for misalignment errors.

- *Casein Glue*—This glue, with gap-filling properties, works well when temperatures are low. It also works well in bonding oily woods such as teak, which could create problems for other glues.

- *Hot-Melt Glue*—This is another relative newcomer on the market. It is applied with a gun. Solid sticks of glue are inserted in a chamber in the gun where they are heated to a liquid state. The glue can then be extruded on the wood. The great advantage of hot-melt glue is that no clamping is required. Just push the items together, and the bond occurs within 30 to 60 seconds, depending on the brand of gun you buy.

- *Epoxy*—Epoxy comes in two parts that are mixed together before use. It can be used to bond china, glass, and a variety of other materials, including wood. For woodwork, use epoxy glue formulated especially for use with wood.

Whatever glue you use, the parts to be mated should be clean, and the glue should be applied as recommended by the manufacturer. Good pressure must be applied, and the mating parts must be in solid contact.

Beveled Joints

In this type of joint, the sides of the pieces fit together to form angles (or corners) as shown in Figure 15-1. An infinite amount of planing and dressing can be saved by first ripping the edges roughly. This can be done by hand, of course, but you may want to take advantage of a table saw. If ripped on a power saw, the angle can be adjusted

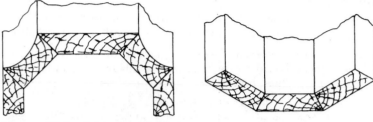

Figure 15-1 Typical beveled joints.

precisely, and only a small amount of hand dressing will be necessary. Try the bevel continually with a T-bevel while dressing to ensure that the joints fit. They must fit properly if the joint is to be glued. The joints must be clamped, and without special clamps, this is troublesome. The woodworker's ingenuity will usually suggest a method. Short pieces of chain with bolts through the end links are useful, if the chains can be passed around the work. Beveled joints do not usually require exceedingly high pressure.

Hidden Slot Screwed Joints

This joint is not often used as a glued joint but is an effective way of fastening brackets and shelves to finished work where the fastening must be concealed. The joint consists of a screw that is driven part way into one piece and a hole and slot cut into the opposite piece. The joint is made by fitting them together with the head of the screw in the hole, which forces the screw back into the slot (see Figure 15-2).

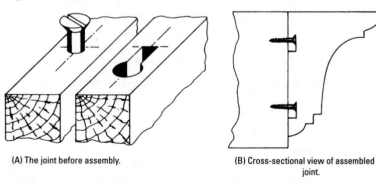

(A) The joint before assembly.

(B) Cross-sectional view of assembled joint.

Figure 15-2 The hidden slot screwed joint.

This joint is also used in interior work for fastening pilasters and fireplaces to walls, for paneling, and for almost every kind of work requiring secure and concealed fastening.

To make the joint, gage a centerline on each of the pieces. Determine the position of the screws and insert them (they should project approximately $^3/_8$ inch above the surface). Hold the two pieces evenly together, and, with a try square, draw a line from the back of the screw shank across the centerline of the opposite piece. From this line, measure $^7/_8$ inch forward on the centerline. With this point as the center, bore a hole to fit the screw head that is just slightly deeper than the amount that the screw projects above the surface. Cut a slot from the hole back to the line from the screw shank. This slot should be as wide as the diameter of the shank and as deep as the hole. As a rule, the total length of the slot and hole should be slightly more than twice the diameter of the head of the screw.

The process of fastening pilasters to fireplaces by this method is as follows. First, mark the position of the piece and the place on the wall for the screws. In brick and cement walls, holes are drilled and wooden plugs are driven in flush with the surface to hold the screws. Plugs shaped as shown in Figure 15-3 seldom work loose. Turn the screws into the plugs and allow them to project approximately $^3/_8$ inch from the surface. The screw heads should be smeared with a bit of lampblack. Put the piece in position and press it against the screw heads. This pressure will leave black impressions. Bore the holes to fit the screw heads approximately $^5/_8$ inch below the impressions, and cut the slot to receive the shank of the screw. Replace the piece with the heads in the holes, and force it down.

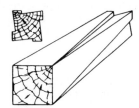

Figure 15-3 A plug is used to hold the screw when hidden slot screwed joints are utilized on mortar and brick walls.

Coopered Joints

These are so named because of the resemblance to the joints used in barrels made by coopers, and are used for practically all forms of curved work. They are usually splined before gluing, although dowels are used occasionally. (Splined and dowel joints are discussed later in this chapter.) Figure 15-4 shows the coopered joint in semicircular form with the segments beveled at an angle of 15 degrees. They are clamped after gluing and planed to shape.

Plain or Butt Joint

The *plain edge joint* is a joint between the edges of boards where the side of one piece is placed against the side of another. The *butt joint* is a joint in which the square end of one member is placed against the square end of another.

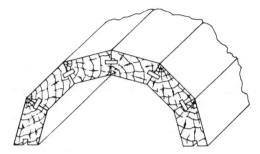

Figure 15-4
Coopered joints are employed to form various curvatures in cabinetwork.

Straight Plain Edge Joint

This type of joint is more or less readily made on a power jointer. The plain edge joint has many uses, and it is commonly used to build up wide boards for panels, shelves, and so on, from narrower pieces. For boat planking, the boards are often curved and slightly beveled so that the joint is left open to be caulked later. Such curved joints must be fitted one edge to the other. In furniture, cabinet, and other fine finish work, the edges are usually glued (see Figure 15-5).

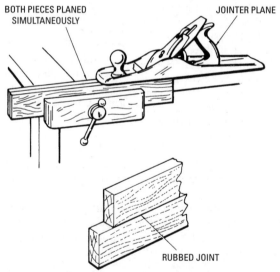

BOTH PIECES PLANED
SIMULTANEOUSLY

JOINTER PLANE

RUBBED JOINT

Figure 15-5 A method of planing both edges together to obtain a straight-butt side joint. This requires a great deal of skill in planing, and it is necessary that the plane be straight on the edge and carefully sharpened and adjusted. After planing, the edges are glued and rubbed together to bond securely.

To make a glued edge joint, square and straighten the edges carefully with planes. Test the edges often with the try square to ensure squareness. To join edges, aliphatic resin (yellow) glues can be used. They dry quickly, are nonstaining, and no special equipment is needed for their application. Yellow glue is not waterproof. If the work is to be used outside, use resorcinol or epoxy adhesives. They are waterproof and can be painted over readily.

All modern glues will function at room temperature. If several boards are to be joined edge to edge (see Figure 15-6), at least three clamps will usually be necessary (one on one side and two on the other side) to prevent buckling. Take care to make the edges true and even when gluing, or it may unnecessarily require considerable planing and scraping to make the joint flush and smooth.

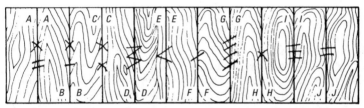

Figure 15-6 Narrow boards can be jointed and placed together by using a marking system so that the same edges will come together when assembling them.

Dowel Joints

There are many variations of dowel jointing commonly used in cabinetwork. It is usually not necessary to dowel a well-fitting glued edge joint, but it is sometimes done to facilitate assembly. The dowels used are usually quite short. For a butt joint into side wood, they are a satisfactory substitute for mortise and tenon joints and are considerably easier to make. When making heavy screen frames, storm sash, and so on, dowel joints are satisfactory if they are glued together with waterproof glue. Figure 15-7 shows the assembly of a typical dowel joint.

Dowels are glued into one piece, cut to length, and sharpened with a dowel sharpener. As a precaution against splitting the joint, cut a V-shaped groove down the side of the dowel with a chisel. This groove permits the glue and air to escape.

The holes must be accurately marked and bored. If these precautions are not taken, the holes will not be in perfect alignment and it will be impossible to assemble the joint or, when assembled, the pieces will not be in their proper alignment. Jigs designed to hold the

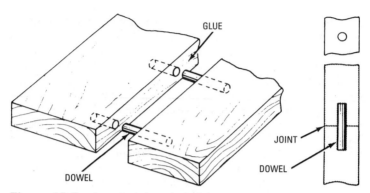

Figure 15-7 A typical dowel pin joint.

bit in alignment are obtainable from several major tool companies, and these devices are a great help when a great amount of doweling is to be done.

The method of making dowel joints without a jig is shown in Figure 15-8. Dowel rods made of several different types of hardwoods are obtainable; some of them have shallow spiral grooves around them to assist carrying the glue into the hole.

To fix the position of dowels accurately in a butt joint (see Figure 15-9), make all measurements and gage lines from the edge of the faced sides. For example, with material that is 4 inches square, mark diagonal lines from the corners with a scratch awl, intersecting at the center. Now, from the edge of the faced sides mark off 1 inch and 3 inches (see Figure 15-10A). From the same sides, gage the lines (see Figure 15-10B). The intersections of the lines are the centers for the holes (see Figure 15-10C).

Where the ordinary means of aligning dowel holes cannot be used, a dowel template or pattern is used. The template is usually made of a strip of zinc or plywood with a small block of wood fastened to one end to act as a shoulder. The position of the dowel point is then pierced through the zinc or plywood pattern with a fine awl. Various types of templates are made and used as the occasion requires. Figure 15-11A shows a template that is used for making dowel rails in furniture. It is made to fit the section of the rail in Figure 15-11B. While held in position, a line is gaged down through the middle. The position of the dowels is indicated on the line. The template is laid flat on a board, and the dowel points are pierced through the surface with a fine awl. When in use, the template is placed in position on the piece, and the dowel positions are marked with an awl through the holes, (see Figure 15-11C). A bit gage should be used

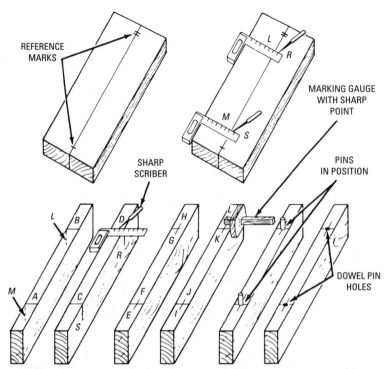

Figure 15-8 Method of making dowel joints. After making reference marks on the two boards, scribe lines A, B, C, and D. Set the marking gauge to half the thickness of the boards, and scribe lines EF, GH, IJ, and KL. Bore a hole at the intersection of each of these lines. The holes should be just less than half the thickness of the boards. The dowels should fit tightly in these holes.

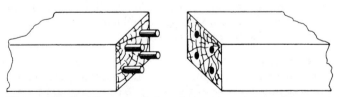

Figure 15-9 The use of dowels in a butt joint adds strength to the joint. This type of construction is frequently found in cabinetwork to lengthen large moldings and where the cross grain prevents tenoning.

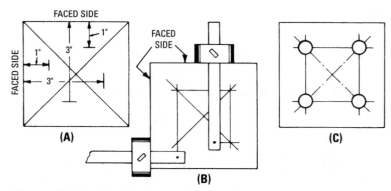

Figure 15-10 The method used for marking the position of the dowels in a butt joint.

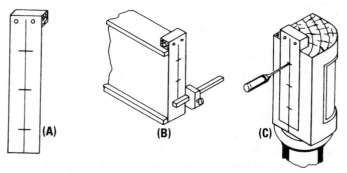

Figure 15-11 The use of a template, or pattern, for marking dowel-pin locations. (A) The template; (B) the template is made to fit the section of rail; and (C) marking dowel positions on leg.

to regulate the depth of the bore when doweling. If a great amount of doweling is to be done, a doweling jig (which ensures accurate boring of holes from $1/4$ to $3/4$ inch) is useful.

Square Corner Joint

The two members of a corner joint are joined at right angles, the end of one butting against the side of the other. When making a corner joint, saw to the squared line with a backsaw and finish with a block plane to fit. The work should be frequently tested with a try square, both lengthwise and across the joint. The method of marking this type of joint is shown in Figure 15-12. The joint may be fastened

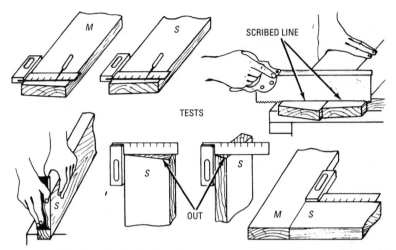

TESTS

SCRIBED LINE

OUT

Figure 15-12 Method of making a corner joint. After squaring and sawing the edges (M and S), plane the joint surface of one board (S) and test the edges with the square until a perfect right-angle fit is obtained.

together with nails or screws. When fastening the joint, the pieces should be firmly held in position at a 90° angle by a vise or by some other suitable means.

Mitered Joints

Mitering is an important part of cabinetwork in the framing of furniture and in panelling, where many difficult moldings must be mitered into place. A plain miter has a spline inserted at right angles to the miter (see Figure 15-13A). This joint is principally used for mitering

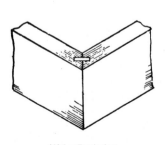

(A) A splined miter.

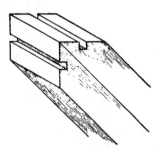

(B) Method of grooving a miter.

Figure 15-13 Mitered joints.

end grain and is additionally strengthened by gluing a block to the internal angle.

One practical way of grooving this joint is to fasten two miters together in a vise so as to form a right angle (see Figure 15-13B), thus providing an edge from which to gage the position of the tongues and plow the grooves. If the pieces are less than 6 inches in width, the grooves are cut with a dovetail saw and chiseled to depth. The grooves can also be cut with a power router or dado blade in a table saw.

A miter joint is used mostly in making picture frames. To properly make a miter joint, a picture-frame vise should be used when fastening the pieces together instead of the makeshift method of offset nailing. In fact, a picture-framing shop, to be worthy of the name, should be provided with a picture-frame vise (see Figure 15-14).

Figure 15-14 A typical picture-frame vise. With this tool, any frame can be held in the proper position for nailing.

Figure 15-15A illustrates frame mitering. It is mostly used for end-grain jointing. For this joint, the tongue should be approximately a third of the thickness of the material, and it may extend all the way through or only part way. This joint is especially useful in cabinet-work for connecting and mitering various types of large moldings around the tops of pieces, for mitering material for tops and panels, and for connecting sections in curved work (see Figure 15-15C).

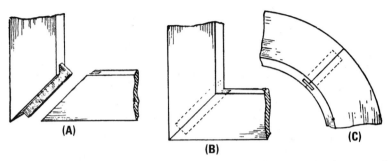

(A) **(B)** **(C)**

Figure 15-15 The frame miter joint in various stages and types of construction. (A) Before completion; (B) completed joint; and (C) for connecting segments in curved work.

Constructing Mitered Joints

When cutting the 45° miter, use a miter box. After sawing, dress and fit the ends with a block plane. There are two ways to nail a miter joint: the correct way with a picture-frame vise and the wrong way with an ordinary vise. Where considerable work is to be done, a combined miter box and vise is desirable (see Figure 15-16).

Figure 15-16 A typical miter machine. With this device, any type of miter joint can be cut, glued, and nailed to make tight, close-fitting corners.

The methods of mitering corners (see Figure 15-17) are used when constructing small drawers for merchandise cabinets (such as

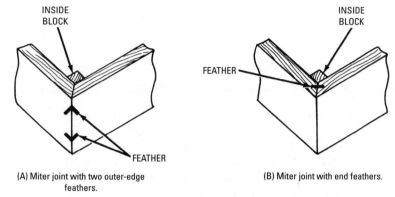

(A) Miter joint with two outer-edge feathers.

(B) Miter joint with end feathers.

Figure 15-17   Miter joints reinforced by feathers. These feathers are kept in place by glue. The joint may also be reinforced by an inside block, as shown.

those used in drug stores). The joinery shown in Figure 15-17A is not particularly effective. A much stronger joint may be made by sawing the groove for the feather straight across the corner almost through. Then glue in strong hardwood feathers, with their ends cut off and smoothed flush. The method shown in Figure 15-17B is not too efficient and is difficult to make because the outside corner of the groove is often chipped in construction. Making the joint can be made easier by sawing the grooves on a band saw (or on a jigsaw, if there are many to make) and then driving a small patented metal feather with sharp turned edges and a slight taper in from each edge. The feathers draw the joint tight, hold well, and no blocking is necessary.

Screwed Miter Joint
Figure 15-18A illustrates a plain miter with a screw driven at right angles to the miter across the joint through a notch cut in the outside of the frame. This type of joint is used principally in light molded frames.

A common method of clamping a tongued miter is to glue blocks to the piece and hand-screw the joint together, as shown in Figure 15-18B. The blocks are glued on and allowed to dry before gluing up the joint. When the joint is dry, the blocks are chiseled off, and any traces of glue sanded out.

Splined Joint
The form of joint shown in Figure 15-19 is called a *splined-joint* (or sometimes a *slip-tongue joint*). In the shop, it is often used for

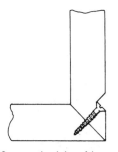

(A) Cross-sectional view of the completed joint.

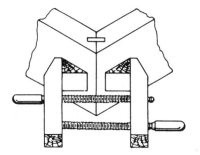

(B) Method of clamping a miter joint.

Figure 15-18   The screwed miter joint.

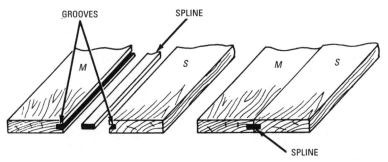

Figure 15-19   The component parts and assembly of a splined joint. The spline fits into the grooves in M and S.

edge-glued joints, since it holds the members in alignment when clamped.

A spline joint is stronger than a simple butt joint (see Figure 15-20A). When the thickness of the material will permit, two splines are used instead of one (see Figure 15-20B) because of the additional gluing surface afforded and the increased strength to the joint.

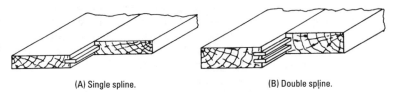

(A) Single spline.

(B) Double spline.

Figure 15-20   Single- and double-splined joints.

To make this joint successfully, the pieces should be properly faced and the edges should be squared and straightened with a jointer so that they fit perfectly. Put reference marks on the face side so that the same edges will come together when assembled. Set the plow plane with the iron projecting approximately $1/32$ inch below the bottom plate. Set the depth gage to half the width of the spline, and adjust the fence so that the cutter will be the required distance from the edge between the two sides. Fasten the piece securely in the bench vise so that the groove can be plowed from the face side. Begin plowing at the front (see Figure 15-21A) and work backward; finish by going right through from back to front, as in Figure 15-21B. Hold the plow plane steady. Otherwise, an irregular groove will result. Grooves can also be cut with an electric router or dado blade on a table saw.

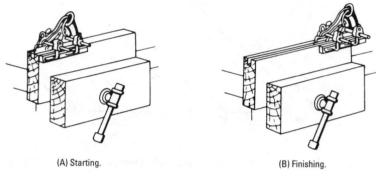

(A) Starting.　　　　　　　　　　(B) Finishing.

Figure 15-21　Method of plowing a single spline groove.

Splines are usually cut with the grain, but cross-grain splines are stronger. For the cross-grain spline, cut off the end of a thin board of hardwood. Mark it, and carefully saw off a strip across its width that is the required width of the spline, approximately $3/4$ inches wide. Plane the spline to the desired thickness in a tonguing board. Then, assemble the parts and glue them up (see Figure 15-22).

A groove is made in each of the pieces to be joined, and a spline (made as a separate piece) is inserted in both grooves. The main reason for the use of a splined joint is that when two pieces of softwood are joined, a hardwood spline (that should be cut across the grain) will make the joint less likely to snap than if a tongue were cut in the softwood lengthwise with the grain.

Splice Joints

A *splice joint* can be used to join pieces of wood end to end. They are joined by fish plates placed on each side and are secured by

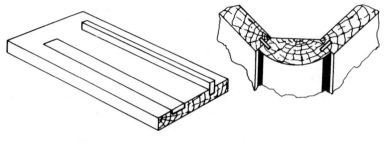

(A) A tonguing board.   (B) Plowed-and-splined joints.

Figure 15-22 The tonguing board is a simple and handy device when used to overcome the difficulty of holding a narrow piece of thin material steady while planing. To make the board, use a piece of $\frac{7}{8}$-inch faced material, 8 to 10 inches wide and longer than the tongues to be planed. Cut the grooves, as indicated, with a tenon saw. Clean out the grooves with a chisel and a router. The wider groove should be slightly deeper than the thickness of the finished tongue to allow for planing both sides of boards placed in it.

cross bolts (or nails) as shown in Figure 15-23. These fish plates may be made either of wood or of iron and they may have plain or projecting ends.

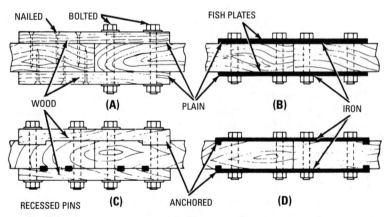

Figure 15-23 Various splice, or fish, joints: (A) plain joint with wooden fish plates; (B) plain joint with iron fish plates; (C) wood plates anchored on the end; (D) iron plates anchored on the end.

The plain type (Figure 15-23A) is normally suitable when the form of stress is compression only. However, if the joint is properly made, it will withstand either tension or compression. If the joint is to be subjected to tension, the fishplates (either wood or iron) should be anchored to the main members by keys or projections, (see Figures 15-23C and 15-23D).

Lap Joints

In the various joints grouped under this classification, one of the pieces to be joined laps over (or into) the other (hence the name *lap joint*). Figure 15-24 shows some typical lap joints.

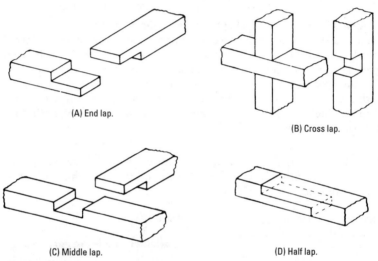

(A) End lap.

(B) Cross lap.

(C) Middle lap.

(D) Half lap.

Figure 15-24 Typical lap joints. The overlapping feature furnishes a greater holding area in the joint and is, therefore, stronger than any of the butt or plain joints. A half-lap joint, sometimes called a scarf joint, is made by tapering or notching the sides or ends of two members so that they overlap to form one continuous piece without an increase in thickness. The joints are usually fastened with plates, screws, or nails and are strengthened with glue.

Rabbet Joint

A *rabbet joint* is cut across the edge or end of a piece of stock. The joint is cut to a depth of about half the thickness of the material (see Figure 15-25). The rabbet joint is a common joint in the construction of cabinets and furniture because it allows pieces to be joined so that no seam shows. On a typical cabinet, for example, the recess (or L-shaped section) of the joint can be cut out of the end of each side and

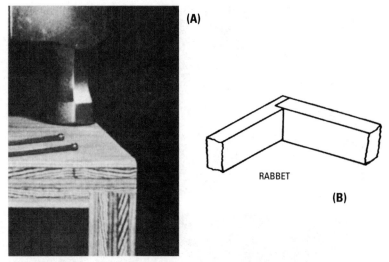

(A)

RABBET

(B)

Figure 15-25 (A) Rabbet joint. An L-shaped section is cut out of the end of one board and the other board is slipped in to the recess, then nailed, screwed, glued, or otherwise attached. (B) Another view of the rabbet joint. *(Courtesy of The American Plywood Assn.)*

the back of the cabinet simply fitted in between the sides and fastened in place. When viewed from the sides, the seam cannot be seen.

Rabbet joints are also used in the construction of drawers, and in the making of boxes of various kinds. A rabbet joint can be made with a wide variety of tools.

Dado Joint

A *dado joint* is a groove cut across the grain that will receive the butt end of a piece of stock. The dado is cut to the width of the stock that will fit in it and to a depth of a third to a half of the thickness of the material (see Figure 15-26). The joint is a common one in construction and is used for the installation of shelves, stairs, and kitchen cabinets.

Scarf Joints

By definition, a *scarf joint* is made by cutting away the ends of two pieces of timber and by chamfering, halving, notching, or sloping, making them fit each other without increasing the thickness at the splice. They may be held in place by gluing, bolting, plating, or strapping.

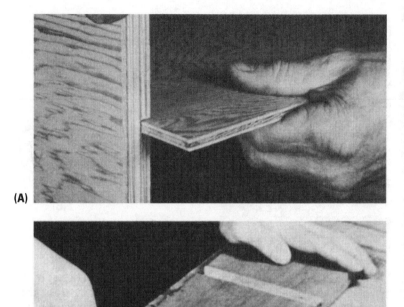

(A)

(B)

Figure 15-26 (A) dado joint. This is a groove cut across the grain, the groove being the width of the material. The dado is a very popular joint for shelves. (B) Another view of dado joint. Here, a Surform tool is being used to clean out dadoes made on the upright member of a shelf unit.

(Courtesy of The American Plywood Assn.)

(Courtesy of Stanley)

There are various forms of scarf joints, and they may be classified according to the nature of the stresses that they are designed to resist as follows:

- Compression
- Tension

- Bending
- Compression and tension
- Tension and bending

Compression Scarf Joint

This is the simplest form of scarf joint. As usually made, one-half of the wood is cut away from the end of each piece for a distance equal to the lap (see Figure 15-27A). This process is called *halving*. The length of the lap should be five to six times the thickness of the timber. Mitered ends (see Figure 15-27B) are better than square ends, where nails or screws are depended on to fasten the joint. For extra heavy-duty joints, iron fishplates are sometimes provided, thereby greatly strengthening the joint (see Figure 15-27C). When these are used, mitered ends are not necessary.

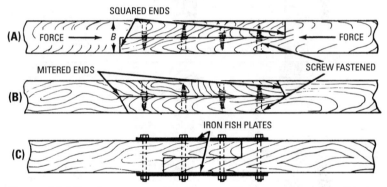

Figure 15-27 Compression scarf joints. (A) Plain square ends; (B) plain mitered ends; and (C) plain square ends, reinforced with iron fishplates.

Tension Scarf Joint

Tension scarf joints used to be made with various methods of locking joints to resist tension (such as by means of keys, wedges, or keys or fishplates with fingers, and so on, as shown in Figures 15-28 and 15-29). Today, you will still see fishplates. They are used, but the keys and wedges are not used in modern joinery (see Figure 15-30).

Bending Scarf Joint

When a beam is acted on by a transverse (or bending stress), the side that the bending force is applied is subjected to a compression stress, and the opposite side is subjected to a tension stress. Thus, in Figure 15-31A, the upper side is in compression, and the lower side is in tension. At L, the end of the joint may be square, but at F, it

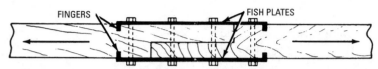

Figure 15-28 Square ends bolted and reinforced with iron fishplates. Tension stress caused by fingers on the fishplates.

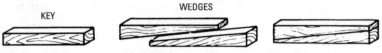

Figure 15-29 Key and wedges.

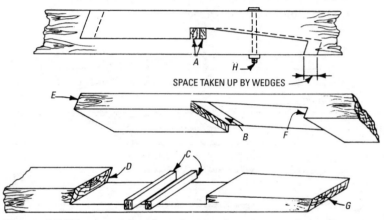

Figure 15-30 A scarf joint with a notch and a mitered half lap. The ends are also mitered, illustrating the location and effect of the wedges. The two pieces are joined together with the wedges (A) driven home and cut off. The dotted lines represent the amount of space closed when the pieces are drawn into place by the wedges. Since the cut is mitered at D, E, F, and G, these boards will form a rigid joint, which is often strengthened by a bolt through each section (H).

should be mitered. If this end were square (as at F′, Figure 15-31B), the portion of the lap M between the bolt and F′ would be rendered useless to resist the bending force.

When designing a bending scarf joint, it is important that the thickness at the mitered end be ample, otherwise the strain applied at the point might split the support. Gluing normally helps prevent such stresses from developing.

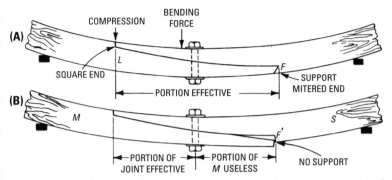

Figure 15-31 Bending scarf joints. One end of the joint should be mitered to provide adequate support for the various stresses applied to the joint.

Mortise-and-Tenon Joints

A *mortise* is defined as a space hollowed out in a member to receive a tenon, and a *tenon* is defined as a projection (usually with a rectangular cross-section) at the end of a piece of member that is to be inserted into a socket (or mortise) in another timber to make a joint.

Mortise-and-tenon joints are frequently called simply *tenon joints*. The operation of making mortise-and-tenon joints is also termed *tenoning* (which also implies mortising).

There are many different mortise-and-tenon joints, and they may be classified with respect to the following:

- Shape of the mortise
- Position of the tenon
- Degree to which tenon projects into mortised member
- Degree of mortise housing
- Number of tenons
- Shape of tenon shoulders
- Method of fastening the tenon

Many variations of the mortise-and-tenon joints are used in cabinetwork. They differ in size and shape according to the requirements of the location and the purpose for which the joint is used (see Figure 15-32).

The most frequently used mortise-and-tenon joint is the stub tenon, so called because it is short and penetrates only part way through the wood (see Figure 15-33). Figure 15-33B shows the stub

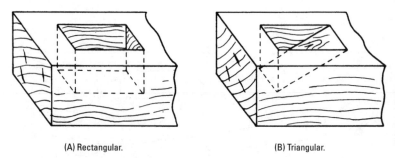

(A) Rectangular.   (B) Triangular.

Figure 15-32 Mortise-and-tenon joints—shape of the mortise.

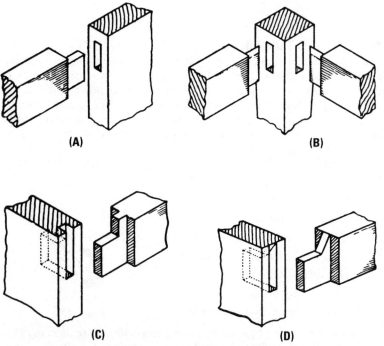

(A)   **(B)**

(C)   **(D)**

Figure 15-33 Variations of the mortise-and-tenon joints are frequently used in doors and in the framing of furniture. (A) Stub mortise-and-tenon; (B) mitered stub tenon; (C) haunched, or rabbeted, mortise-and-tenon; and (D) haunched and mitered mortise-and-tenon.

tenon with a mitered end—this type of construction is often necessary when fitting rails into a corner post. Figure 15-33C shows the *rabbeted* or *haunched tenon*, which is considered a stronger joint because of the small additional tenon formed by the rabbet. It is often mitered, as shown in Figure 15-33D, to conceal the joint when used on outside frames.

The joint in Figure 15-34A is the same as that shown in Figure 15-33C, but it is shouldered on one side only. It is sometimes called a barefaced tenon and is used when the connecting rail is thinner than the stile into which it is joined. Figure 15-34B shows the long

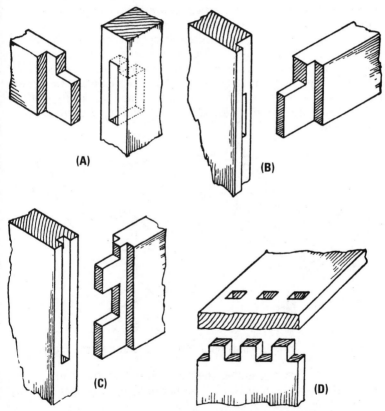

(A)

(B)

(C)

(D)

Figure 15-34 Other mortise-and-tenon joints used in the construction of furniture. (A) Barefaced mortise-and-tenon; (B) long and short shoulder mortise-and-tenon; (C) double mortise-and-tenon; and (D) pinned mortise-and-tenon.

and short shoulder tenon. This joint is used when connecting a rail into a rabbeted frame, since it has one shoulder cut back to fit into the rabbet. Figure 15-34C illustrates the double tenon joint that increases the lateral strength of the stile into which it is jointed. A stub-tenon joint is rabbeted and notched to form two tenons. When glued, it makes an exceptionally strong joint. Figure 15-34D represents a type of through mortise-and-tenon that is sometimes used for mortising partitions into the top or bottom of wardrobes, cabinets, and so on. The partitions are wedged across the tenon and glued.

The mortise and tenon must exactly correspond in size (that is, the tenon must fit into the mortise accurately). The position of the tenon is usually at the center of the member, but sometimes it is located at the side, depending (except in special cases) on the degree of housing (see Figure 15-35). The tenon may project partly into (or through) the mortised member. When the tenon and mortise do not extend through the mortised member, the joint is called a stub tenon. This form of tenon is used for jointing the framework of partitions and is employed in work where the joint will not be subjected to any tension.

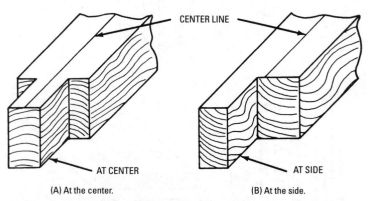

Figure 15-35 Mortise-and-tenon joints—position of the tenon.

Degree of housing signifies the degree to which the tenon is covered by the mortise, that is, the number of sides of the mortise (see Figures 15-36 and 15-37).

The number of tenons depends on the shape of the members, whether they are square or rectangular, with considerable width and little thickness (see Figure 15-38). The tenon shoulders are usually at right angles with the tenon as they are when the two members

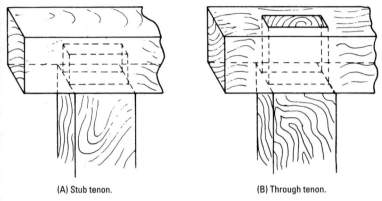

(A) Stub tenon.   (B) Through tenon.

Figure 15-36   Mortise-and-tenon joints—degree to which the tenon projects into the mortised timber.

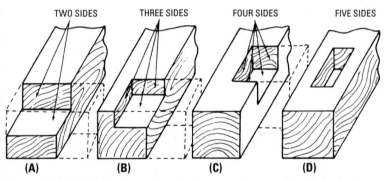

TWO SIDES   THREE SIDES   FOUR SIDES   FIVE SIDES

(A)   (B)   (C)   (D)

Figure 15-37   Mortise-and-tenon joints—degree of mortise housing: (A) two sides; (B) three sides; (C) four sides; and (D) five sides.

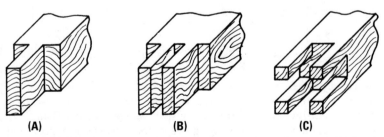

(A)   (B)   (C)

Figure 15-38   Mortise-and-tenon joint—number of tenons. (A) Single tenon; (B) double tenon; and (C) multi-tenon. The double and multitenon have not been used for many years.

are joined at right angles, but they may be mitered to some smaller angle, such as 60° or 45°, as in the case of a brace (see Figure 15-39).

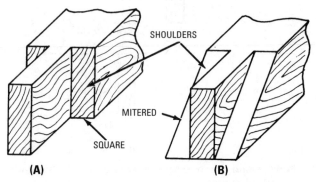

Figure 15-39 Mortise-and-tenon joints—shape of the tenon shoulders. (A) square and (B) mitered.

There are several ways of fastening mortise-and-tenon joints, such as with pins or wedges (see Figure 15-40). When making a mortise-and-tenon joint, the work is first laid out to given dimensions.

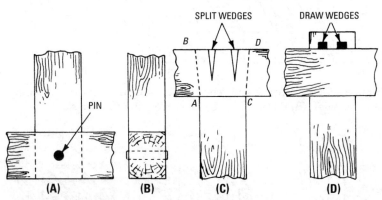

Figure 15-40 Shown are former methods of fastening the tenon: (A) side view, tenon secured by a pin; (B) front view, tenon secured by a pin; (C) tenon secured by internal, or split, wedges (sides AB and CD are tapered, thus securely wedging the tenon into the mortise); and (D) tenon secured by external, or draw, wedges, which are driven into rectangular holes beyond the mortise. Today, strong glue and proper clamping usually do the job.

Laying Out the Mortise-and-Tenon

The general practice when laying out a mortise-and-tenon is to square the mortise lines across the edge of the stile in pencil and then scribe two lines for the sides of the mortise with a mortise or slide gage between the pencil lines (see Figure 15-41). If the tenon is

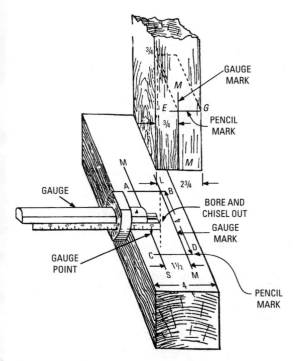

Figure 15-41 Method of laying out a mortise-and-tenon joint having mortise in the center of one timber and tenon at the side of the other timber. Set gauge to one-third the thickness of the timber to be mortised, then with gauge scribe line MS, and with gauge on opposite side of the timber, scribe line LF. These lines define the sides of the mortise. To complete the mortise lay out lines AB and CD, using square and pencil. Without changing the setting on the gauge, scribe line M'S', for tenon on the other timber and EG, with square at correct distance from the end. In working on finished surfaces be careful not to scribe the lines MS and LF beyond the ends of the mortise as the scratches will show after the job is completed. In such cases, it is better to mark lightly with pencil AB and CD first.

to be less than the full width of the rail, square the rail lines across the edge in addition to the mortise lines (see Figure 15-42A). This procedure ensures greater accuracy when designating the position of the mortise. When two or more stiles are to be mortised, they are clamped together, and the lines are squared across all the edges simultaneously.

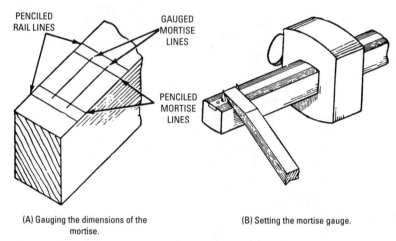

(A) Gauging the dimensions of the mortise.

(B) Setting the mortise gauge.

Figure 15-42 Laying out the mortise.

For a through mortise, continue the pencil lines across the face side and onto the back edge. Gage the mortise lines from the faced side. With the gage set for the mortise scribe the lines for the tenon on both edges and the end of the rail, and with the aid of a try square, mark the shoulder lines with a knife or chisel on all four sides.

The proportions of stub and through mortises-and-tenons are usually about a third of the thickness of the wood, and they should be cut with a sharp chisel of the required size. If the chisel is not exactly a third of the thickness of the material, it is better to make the mortise more than a third, rather than less. Set the mortise gage so that the chisel fits exactly between the points (see Figure 15-42B). Make a chisel mark in the center of the edge to be mortised, and adjust the head of the gage so that the points coincide with this mark.

Mortise cutting in cabinetwork can be done entirely with a sharp chisel, beginning at the center and working toward the near end with the flat side of the chisel toward the end. Remove the core as you proceed, then reverse the chisel and cut to the far end, being careful to keep the chisel in a perpendicular position when cutting the ends. Through mortises are cut halfway through from one side and the material is then removed and cut through from the opposite side.

A depth gage for stub mortises is made by gluing a piece of paper or tape on the side of the chisel (see Figure 15-43). If the method of boring a hole in the center from which to begin the cutting of the mortise is used for stub mortises, it is advisable to use a bit gage to regulate the depth of the bore. A small firmer chisel is used to clean out stub mortises.

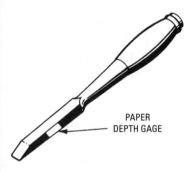

Figure 15-43 The depth of the mortise joint may be controlled by fastening a piece of paper or tape on the side of the chisel.

PAPER
DEPTH GAGE

Cutting the Mortise

Select a chisel that is as near to the width of the mortise as possible. This chisel (especially for large work) should be a framing or mortise chisel. Bore a hole the same size as the width of the mortise at the middle point. If the mortise is for a through tenon, bore halfway through from each side. In the case of a large mortise, most of the wood may be removed by boring several holes (Figure 15-44). When cutting out a small mortise with a narrow chisel, work from the hole in the center to each end of the mortise, holding the chisel firmly at right angles with the grain of the wood. At the ends of the mortise, the chisel must be held in a vertical position (see Figure 15-45B) with the flat side facing the end of the mortise.

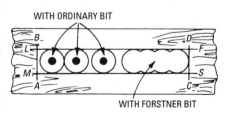

WITH ORDINARY BIT

WITH FORSTNER BIT

Figure 15-44 A method of boring holes when making a large mortise.

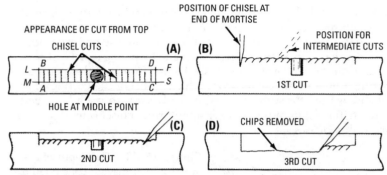

Figure 15-45 A method of cutting a small mortise. After laying out the mortise, bore a hole at the center (A) and work toward each end with a chisel. The chisel cuts should always be made across the grain.

Always loosen the chisel by a backward movement of the handle. A movement in the opposite direction would damage the ends of the mortise. Never make a chisel cut parallel with the grain because the wood at the side of the mortise may split. When cutting a through mortise, cut only halfway through on one side, and finish the cut from the other side. After cutting, test the sides of the mortise by using a try square, as shown in Figure 15-46. This procedure will check the accuracy with which the work was laid out.

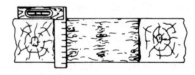

Figure 15-46 Test the end with a square after cutting the mortise.

Cutting the Tenon

A backsaw is used for cutting out the wood on each side of the tenon, and, if necessary, a finishing cut may be taken with a chisel. After the wood has been cut away, the tenon should be pointed by chiseling all four sides.

Fasten the piece firmly in the bench vise. Start the cut on the end grain, and saw diagonally toward the shoulder line (see Figure 15-47A). Finish by removing the material in the vise and cutting downward flush with the edge (see Figure 15-47B). The diagonal saw cut acts as a guide for the finishing cut and provides greater accuracy. Small tenons are usually cut with a dovetail saw.

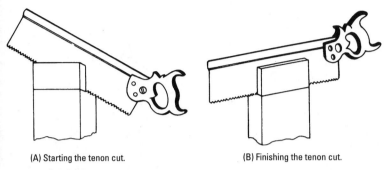

(A) Starting the tenon cut.   (B) Finishing the tenon cut.

Figure 15-47   Making the tenon cut.

Figure 15-48 shows the appearance of the tenon before and after the pointing operation. If this operation were omitted, a tight-fitting tenon would be difficult to start into the mortise and could splinter the sides of the mortise when driven through on a through mortise. Do not cut off the point until the tenon is finally in place and the pin is driven home.

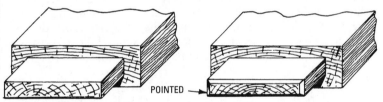

POINTED →

Figure 15-48   Appearance of the tenon before and after pointing.

Cutting the Shoulder

After making the tenon cuts, and to overcome any difficulty in cutting the shoulders, place the piece on the shoulder board or bench hook and carefully chisel a V-shaped cut against the shoulder line (see Figure 15-49A). Hold the work firmly against the stop on the board. Place the saw in the chiseled channel, and begin cutting by drawing the saw backward and then pushing it forward with a light stroke. Hold the thumb and forefinger against the saw (see Figure 15-49B), and keep the saw in an upright position. A straightedge can be placed against the shoulder line to act as a guide when cutting wide shoulders. In the case of extremely wide tenons and shoulders, a rabbet plane and a shoulder plane are used. A straightedge is used as a guide for the rabbet plane.

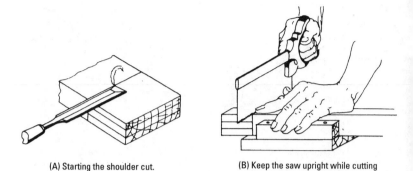

(A) Starting the shoulder cut.

(B) Keep the saw upright while cutting the shoulder.

Figure 15-49 Making the shoulder cut.

Draw Boring

The term *draw boring* signifies the method of locating holes in the mortise and tenon that are eccentric with each other so that when the pin is driven in, it will draw the tenon into the mortise, thereby forcing the tenon shoulders tightly against the mortised member. The holes may be located either by accurately laying out the center (as shown in Figure 15-50C) or by boring the mortise and finding the center for the tenon hole (as in Figure 15-50D). Considerable experience is necessary to properly locate the tenon hole. If too much offset is given, an undue strain will be brought to bear on the joint.

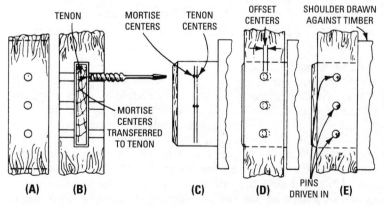

Figure 15-50 A method of transferring pin centers from the mortise holes to the tenon by draw boring. When laying out the tenon-hole centers, make the offset toward the tenon shoulder.

This strain is frequently sufficient to split the joint. It is much better to accurately lay out the work and make a tightly fitting pin than to depend on draw boring.

Dovetail Joints

A *dovetail joint* may be defined as a partially housed, tapered mortise-and-tenon joint, the tapered form of mortise and tenon forming a lock that securely holds the parts together. The word dovetail is used to describe the way the tenon expands toward the tip and resembles the fanlike form of the tail of a dove. The various forms of dovetail points (some of which are shown in Figure 15-51) may be classed as follows:

- Common
- Compound
- Lap or half-blind
- Mortise or blind

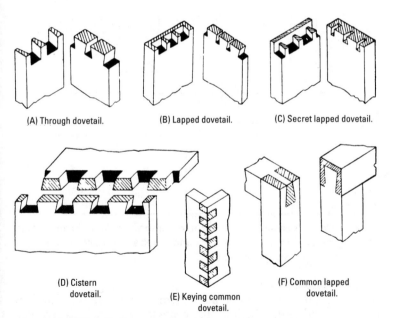

(A) Through dovetail.

(B) Lapped dovetail.

(C) Secret lapped dovetail.

(D) Cistern dovetail.

(E) Keying common dovetail.

(F) Common lapped dovetail.

Figure 15-51 Different types of dovetail joints. Dovetail joints are used principally in cabinetmaking, drawer fronts, and fine furniture work. They are a partly housed and tapered form of tenon joint in that the taper forms a lock to hold the parts securely together.

In common dovetailing, it is a matter of convenience whether to cut the pins or the dovetails first. However, where a number of pieces are to be dovetailed, time can be saved by clamping them together in the vise and cutting the dovetails first.

Common Dovetail Joint

This is a plain or single pin joint. In dovetail joints, the tapered tenon is called the *pin*, and the mortised part that receives this joint is called the *socket*. Where strength rather than appearance is important, the common dovetail joint is used (see Figures 15-52 and 15-53).

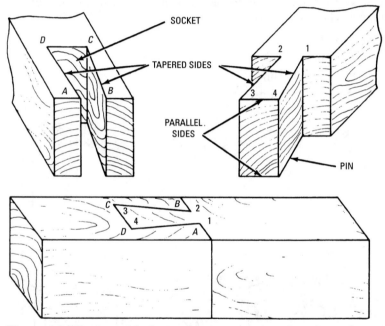

Figure 15-52 A straight form of the common, or plain, dovetail joint. By noting the positions of the letters and numbers, it may be seen how the socket and pin are assembled.

Compound Dovetail Joint

This is the same as the common form but has more than one pin, thereby adapting the joint for use with wide boards (see Figure 15-54). When making this joint, both edges are made true and square. A gauge line is run around one board at a distance from the end equal to the thickness of the other board, and the other board is treated similarly. Two methods are commonly followed. Some mark and cut the pins first; others mark and cut the sockets first.

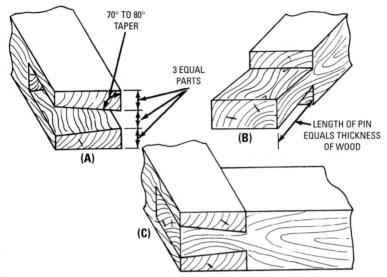

Figure 15-53 A corner form of the common, or plain, dovetail joint with the proper proportions for the socket and pin.

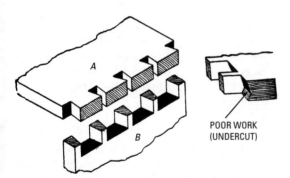

Figure 15-54 A compound dovetail joint, with a poorly cleaned joint shown in detail.

In the first method, the pins are carefully spaced, and the angles of the tapered sides are marked with the bevel. Saw down to the gage line, and work the spaces in between with a chisel and a mallet. Then, put *B* on top of *A* (see Figure 15-54) and scribe the sockets. Square over, cut down to the gage line, clean out, and fit together.

The second method is to first mark the socket on *A* (sometimes on common work, the marking is dispensed with, and the worker uses his eyes as a guide) and second, run the saw down to the gauge line, put *A* on *B*, and mark the pins with the front tooth of the saw. Cut the pins, keeping outside of the saw mark sufficiently to allow the pins to fit tightly. Both pieces may then be cleaned out and tried together.

When cleaning out the sockets and the spaces between the pins, the woodworker must cut halfway through, then turn the board over and finish from the other side, taking care to hold the chisel upright so as not to undercut (see Figure 15-54), which is sometimes done to ensure the joint fitting on the outside.

Lap or Half-Blind Dovetail Joint

This joint is used in the construction of drawers in the best grades of work. The joint is visible on one side but not on the other (see Figure 15-55), hence the name *half-blind*. Since this form of dovetail joint is used so extensively in the manufacture of furniture, machines have been devised for making the joint, thus saving time and labor.

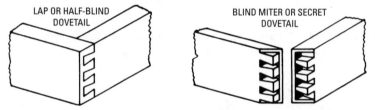

Figure 15-55 Half-blind and blind dovetail joints. These joints were used in the best grades of drawer and cabinetwork because the joint is visible on only one side. They had to be exceptionally well fitted because of the frequent pull on the front piece.

Blind Dovetail Joint

This is a double lap joint (that is, the joint is covered on both sides, as shown in Figure 15-55) and is sometimes called a *secret dovetail joint*. The laps may be either square, as in Figure 15-56, or mitered, as in Figure 15-57. Because of the skill and time required to make these joints, they are used only on the finest work. The mitered form is the more difficult of the two to assemble.

Spacing

The maximum strength would be gained by having the pins and sockets equal. However, this is rarely done in practice because the mortise is made so that the saw will just clear at the narrow side with the space from eight to ten times the width of the widest side. Small pins are used for the sake of appearance, but large ones are

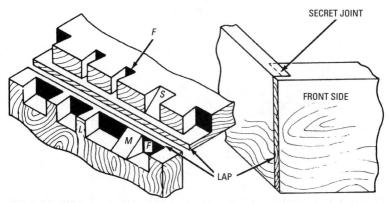

Figure 15-56 A blind, square-lap dovetail joint is another useful joint. Two forms of pins and sockets were used—mitered (MS) and square (LF).

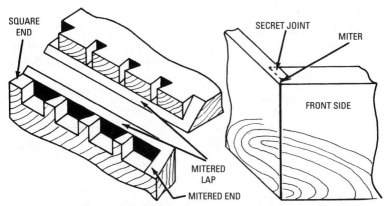

Figure 15-57 A blind, mitered-lap dovetail joint, another oldie.

preferable. The outside pin should be larger than the others and should not be too tight or there will be the danger of splitting, as shown in Figure 15-58 at point *A*. The angle of taper should be slight (70° to 80°) and not acute as shown, otherwise there is the danger of pieces *L* and *F* being split off in assembling.

Position of Pins

When boxes are made, the pins are generally cut on the ends with the sockets on the sides. Drawers have the pins on the front and back. The general rule is to locate the tapered sides so that they are

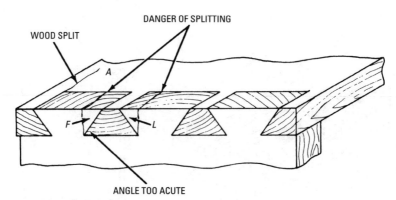

Figure 15-58 A badly proportioned common dovetail joint can result in splitting.

in opposition to the greatest stress that may be applied on the piece of work to which the joint is connected.

Tongue-and-Groove Joint
In this type of joint, the tongue is formed on the edge of one of the pieces to be joined, and the groove is formed in the other, as shown in Figure 15-59.

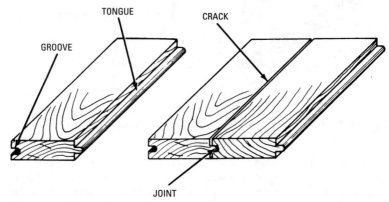

Figure 15-59 The tongue-and-groove joint.

Dovetail Angles
For particular work where the joint is exposed, the dovetails should be cut at an angle of 1 in 8, and for heavier work, 1 in 6. To find

the dovetail angle, draw a line square with the edge of a board, and divide it into 6 or 8 equal parts as desired. From the end of the line and square with it, mark off a space equal to one of the divisions, and set the bevel as shown in Figure 15-60.

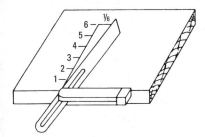

Figure 15-60   Finding the angle of the dovetail.

A dovetail template (see Figure 15-61A) will be quite handy if there is a great deal of dovetailing to be done. To make the template, take a rectangular piece of ³/₄-inch material of any desired size, and square the edges; with the mortise gage set for a ¹/₄-inch mortise at ¹/₄ inch from the edge, scribe both edges and one end. With the bevel set as shown (see Figure 15-60), mark the shoulder lines across both sides of the lower portion, and cut it with a tenon saw. Make one cut for each of the two angles. The template may also be made by gluing a straightedge (at the required angle) across both sides at one end of a straight piece of thin material. The use of a dovetail template saves time and ensures uniformity. Place the shoulder of the template against the edge (see Figure 15-61B) and mark one side of the dovetail along its edge. Reverse the template, place the other shoulder at the same edge, and mark the other side of the dovetail.

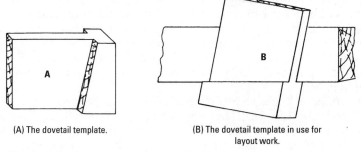

(A) The dovetail template.

(B) The dovetail template in use for layout work.

Figure 15-61   A template is invaluable for dovetail work.

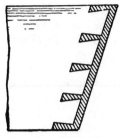

Figure 15-62 Beveled dovetail joint.

Beveled Dovetailing

The joint shown in Figure 15-62 is sometimes required in cabinetwork, and a template can be a great help for marking it. To use the template for marking beveled dovetails, cut a wedge-shaped piece of material (see Figure 15-63A) that is beveled at the same angle as the bevel of the material to be dovetailed. Insert this wedge between the edge of the material and the template, with the square edge of the wedge against the shoulder of the template (see Figure 15-63B). Mark the dovetail as described, but do not reverse the wedge-shaped piece.

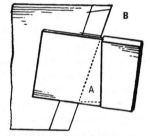

(A) A wedge is used with the dovetail template to mark the desired bevel.

(B) Laying out beveled dovetails with the template and wedge.

Figure 15-63 The method of laying out work with the aid of a dovetail template.

The common, or *through dovetail* (see Figure 15-64A), is primarily used for dovetailing brackets and frames that are subject to a heavy downward strain. Figure 15-64B illustrates the common *lapped* (or *half-blind*) *dovetail* as it is applied to a curved doorframe. It is used in all locations of this type where mortise-and-tenon joints would not be effective. The common lapped dovetail joint may also be used for purposes similar to those described for the common dovetail joint.

The common housed bare face dovetail is shouldered on one side only (see Figure 15-65A). The joint in Figure 15-65B is shouldered and dovetailed on both sides and is another of the same type with the dovetailing parallel along its entire length. These are the simplest forms of housed dovetailing. Their application is usually for the framing of furniture.

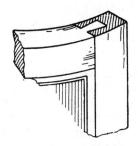

(A) Common dovetail.   (B) Common lapped dovetail.

Figure 15-64 Two typical dovetail joints.

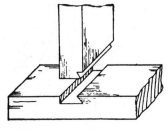

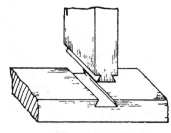

(A) Barefaced dovetail housing.   (B) Common housed dovetail.

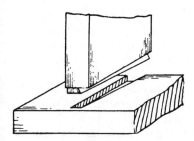

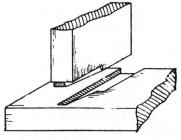

(C) Shouldered housing dovetail.   (D) Dovetailed and housed.

Figure 15-65 Housed dovetails of the single- and double-shouldered types.

Figure 15-65C illustrates a shouldered housing dovetail joint with the dovetail tapering along its length. As with the two preceding joints, this joint can be shouldered on one side or both sides. The tapered dovetail makes this joint particularly adaptable for connecting fixed shelves to partitions because the dovetails prevent the partitions from bending. A dovetailed and housed joint (frequently called a *diminished dovetail*) is principally used on comparatively small work, such as small fixed shelves and drawer rails (see Figure 15-65D).

Making a Diminished Dovetail

Square division lines across the ends into which the shelf is to be housed and dovetailed as far apart as the thickness of the shelf, and gage the depth of the housing on the back edge. Gage lines $3/8$ and $4\frac{1}{2}$ inches from the front edge between the division lines; the space between the gage lines is the length of the actual dovetail (see Figure 15-66A).

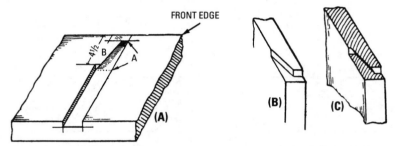

Figure 15-66 The method of making a diminished dovetail joint.

Cut out the section indicated at *A* with a chisel, and undercut side *B* to form a dovetail. Insert a tenon saw, and cut the sides across to the edge. Remove the core with a firmer chisel, and finish to depth with a router. Gage lines on both ends of the shelf on the side and end for the depth of the dovetail, and square across the end the distances from the front edge as given. Cut away the surplus wood with a tenon saw (see Figure 15-66B), and finish the cut with a chisel, carefully testing until it fits hand tight. Figure 15-66C shows the completed end. The average length of the actual dovetail of this type is slightly less than a quarter of the total length of dovetail and housing.

Halved Lap and Bridle Joints

These are lap joints with each of the pieces halved and shouldered on opposite sides so that they fit into each other. Figure 15-67A shows the *common halved angle*, which is the one most frequently used. Figure 15-67C illustrates the *oblique halved joint*, which is used for oblique connections. Figure 15-67D represents the *mitered halved joint*, which is useful when the face or frame piece is molded. Figures 15-67E, F, and G show the joints that are used for cross connections having an outside strain. Figure 15-67H illustrates the

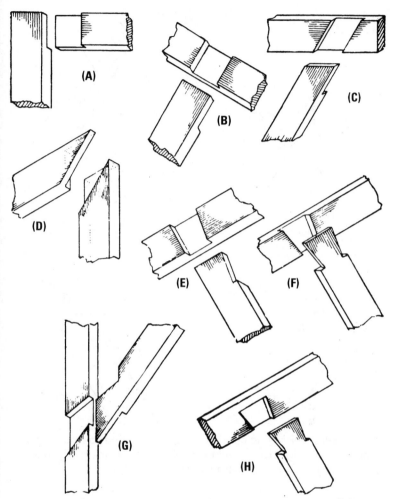

Figure 15-67 Various halved joints: (A) halved lap joint; (B) halved tee joint; (C) oblique halved joint; (D) mitered halved joint; (E) dovetail halved joint; (F) dovetail halved joint; (G) oblique dovetail halved joint; and (H) blind dovetail halved joint.

blind dovetail halved joint, which is used in places where the frame edge is exposed. Bridle or open-tenon joints are used to connect parts of flat and molded frames. The joint in Figure 15-68B is used where a strong, framed groundwork (which is to be faced up) is required. The joint in Figure 15-68D is used as an inside frame connection.

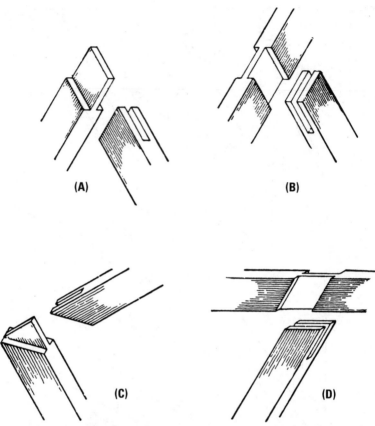

(A)   **(B)**

(C)   **(D)**

Figure 15-68   Halved joints that are generally used to make flat and molded frames: (A) angle bridle joint; (B) tee bridle joint; (C) mitered bridle joint; and (D) oblique bridle joint.

Construction

There are two basic types of construction used in cabinetmaking: framed and carcass. Although the examples to follow show them as separate and distinct types, they are sometimes used in combination, even on a single cabinet.

Framed Construction

In *framed construction*, individual sticks of wood are joined to form a framework (see Figure 15-69). Mortise-and-tenon or doweled joints are commonly used. The outer spaces of the framework might be filled in with panels, to enclose the space within, or with drawers. Horizontal members can be covered to form open shelves or work surfaces.

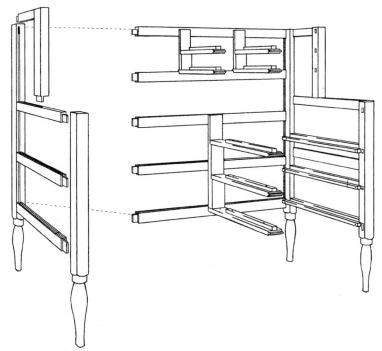

Figure 15-69 Framed construction of chest of drawers. Mortise-and-tenon joinery is used. Tenons are cut on the ends of horizontal rails that fit into mortises cut in the vertical posts.

The joints and framework must be carefully planned and detailed to ensure the accurate fit these types of joints require for maximum strength. A dry assembly is usually done to ensure that all parts are correctly shaped. Then the frame is disassembled. For final assembly the panels are slid into place, glue is applied to the joints, and the framework is clamped.

Carcass Construction
In *carcass construction*, large panels are connected with continuous joints (see Figure 15-70). Splined-miter, housed, and dado joints are commonly used. Carcass construction takes full advantage of large sheets of plywood and other types of manufactured boards.

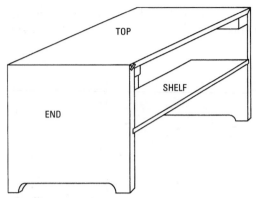

Figure 15-70 Carcass construction of a desk. This back view shows the splined miter joint used to connect the top and end panels. A housed joint cut on the inside of the end panel holds the edge of the shelf.

The process of fabricating panels into carcasses is simple compared to framed construction. Parts are first cut to size. Then joints are cut along edges and within the surface of the panels. Final assembly is usually done in one operation with glue and screws.

Figure 15-71 represents various joints used to connect panels in carcass construction. The joints shown in Figure 15-71A and Figure 15-71B are identical except for the return bead, which is worked on one of the pieces. These two joints are usually glued and nailed. The splined joint Figure 15-71C may be used to connect framing at any angle. The spline prevents slipping during clamping. Figure 15-71D shows an ordinary rabbeted joint with the corner rounded off and the pieces glued together. Because of its rounded corner, it is often used in furniture for children's nurseries. Figure 15-71E illustrates a joint that is shouldered on one side only. A bead is worked on the tongue piece to hide the joint. It is used for both internal and external angles, with or without the bead. The splayed corner tongued joint is used for joining sides into a pilaster corner (see Figure 15-71F).

Plate Jointery

The doweled joints described earlier are easy to make, but even with jigs, alignment is difficult at best. During the 1950s, Swiss cabinetmaker Herman Steiner developed an alternative to doweled joints. Plate jointery substitutes a flat oblong plate of wood for dowels.

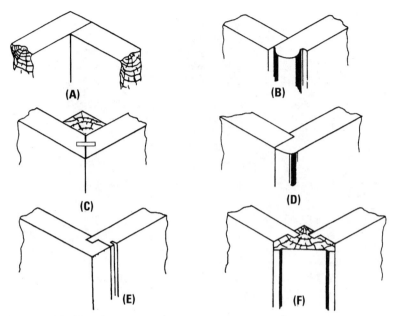

Figure 15-71   Commonly used framing joints: (A) butt or square joint; (B) return bead and butt joint; (C) splined miter joint; (D) rabbet and round joint; (E) barefaced tongued joint; (F) splayed corner joint.

The System
A special hand-held machine is used to cut circular grooves into each member of the joint. Glue is applied to the grooves and special plates inserted. Moisture from the glue swells the plates, producing an exceptionally strong joint.

Layout
Laying out joints is much faster and easier than with doweled joints. Grooves are cut slightly longer than the plates. When the joint is assembled, the parts can shift sideways about 1/8 inch. The machine's faceplate automatically gages the distance of the groove from the face of the work piece. This means that the layout can be quickly done with less attention to exact markings. To lay out for a mid-panel butt joint, simply locate the position of the joint with two measurements and lay the adjoining panel right beside the marks (see Figure 15-72). Then mark the approximate position of each plate

Figure 15-72 Edge grooves are cut in edge of the adjoining panel.
(Courtesy Steiner Lamello AG, Switzerland)

along the joint with a quick pencil mark. The location is usually judged by eye and sometimes location marks for each plate are not needed.

Cutting Grooves

The heart of the system is the special machine used to cut the grooves. It is built with a high-speed electric motor that drives a 4-inch diameter carbide saw blade attached to an adjustable faceplate. In use, the cutter emerges through a slot in the faceplate as it is pressed against the edge of the workpiece. The depth of cut determines one of the three sizes of groove. The machine is portable and typically hand held, which means that it can be easily brought directly to the spot where the joint is needed. To cut the grooves for a mid-panel butt joint, turn on the machine, lay the faceplate on the panel, and nestle it against the edge of the adjoining panel. Press the motor housing against the edge and the cutter emerges within the joint to cut the groove (see Figure 15-72). To cut matching grooves in the face of the panel, orient the machine vertically and repeat the plunge-cut action (see Figures 15-73 and 15-74A). Attachments and fittings for bench mounting and foot actuation are available. Bench mounting would be suitable for high rates of production on many smaller-sized parts.

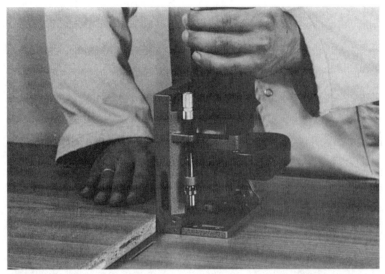

Figure 15-73 Mid-panel grooves are cut in the face of the panel.
(Courtesy Steiner Lamello AG, Switzerland)

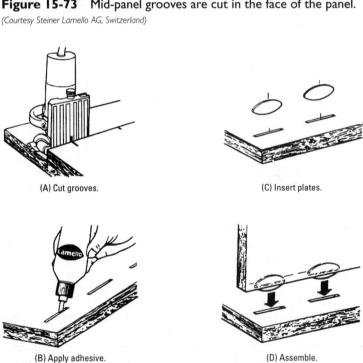

(A) Cut grooves.

(C) Insert plates.

(B) Apply adhesive.

(D) Assemble.

Figure 15-74 The four steps of plate jointery.
(Courtesy Steiner Lamello AG, Switzerland)

Glue Application

A water-based adhesive must be used. Coat all the interior surfaces of each groove (see Figure 15-74B). Squeezable bottles are available with special tips to fit each of the three groove sizes. The tips meter and spread the glue in one quick operation.

Plates

The plates (or *biscuits*) are specially manufactured for this system of jointery. They are supplied in three sizes: #0 ($^5/_8$ inch × $1^3/_4$ inches), #10 ($^3/_4$ inch × $2^1/_8$ inches), and #20 (1 inch × $2^3/_8$ inches). Wood plates are made of solid beech and compressed under great pressure. The result is a typical crosshatch pattern crimped into the surface that makes them look like biscuits.

Metal and plastic plates are also available for special applications. For low-volume custom work, insert plates by hand (see Figure 15-74C). The fit is loose because the plate is compressed. For high-volume production work, hand-held pneumatic installing guns are available. They hold a stack of plates and insert them one at a time into grooves.

Assembly

Once glue is applied and plates are inserted, the assembly must be completed quickly (see Figure 15-74D). The plate begins absorbing moisture from the glue and swelling as soon as it is inserted. Depending on the panel material, glue used, and the air temperature, an average open time is 10 minutes. Once the joint is assembled, check alignment because the oversized slots allow some sideways movement (see Figure 15-75). Complete the assembly and clamping before the plate is locked in place. Frequent clamping in is not needed.

Advantages and Disadvantages

Plate joints are fast to cut and assemble. Other devices such as *knock-down fasteners* and *hinges* can be inserted into plate grooves, expanding the flexibility of the system. There are several manufacturers of both machines and plates, so you can shop to find a system that meets your specific needs and budget.

The adjacent surfaces of a butt joint may not be flush because there is some looseness in the joint until the plate swells. The pressure of the swelling may not be enough to align the surfaces as the glue sets. When substituting plate joints for mortise-and-tenon joints, the rails must be at least $1^3/_4$ inches wide, or the groove will show.

(A) Corner butt joint.

(B) Mid-panel butt joint.

(C) Miter joint.

(D) Edge butt joint.

(E) Frame joint.

Figure 15-75   Common plate joint configurations.

(Courtesy Steiner Lamello AG, Switzerland)

Summary

A great variety of joints used in cabinetwork are usually classified according to their general characteristics (such as glued, halved and bridle, mortise-and-tenon, dovetail, miter, framing, hinging, and shutting).

In cabinetwork, all joints should be glued. Most glued joints must be clamped, or pressure must be applied by means of nails or screws.

The glue that is the most popular in woodworking shops is yellow carpenter's glue. This glue does not stain wood. It is light in color, easy to use, and sets at room temperature. Glued joints must be in close contact while they are setting up. If this contact is not made, the joint will not have the strength required to hold up under normal service.

In carpentry and woodworking, the term joint means the union of two or more members. The goal is to achieve a strong joint without weakening any part of the rest of the structure by removing too much stock.

There are many types of wood joints, which may be divided into classifications such as plain or butt joints, lap joints, mortise-and-tenon joints, and dovetail joints, to name a few. A plain or butt joint is where the end or one side of a piece is placed or butted against one end or side of the other. A lap joint is made when two pieces to be joined lap over or into one another.

Plain or butt joints are generally classified as straight or plain-edge, dowel pin, splined or feather, beveled spline or miter, and beveled plain edge. Straight or plain-edge joints are the simplest form of joint and have many uses where several pieces are required to form a flat surface.

The dowel joint can be considered as a substitute for mortise-and-tenon joints. If well-made and not exposed to weather and extreme temperature changes, it is a strong and excellent joint. A dowel joint is simply a butt joint reinforced by dowels that fit tightly into holes bored in each member to align them with each other.

There are many variations of the mortise-and-tenon joint. Among the more common joints are stub tenon, through tenon, haunched tenon, open tenon, and double tenon. When cutting a mortise, select a chisel as near the width of the mortise as possible. Check all cuts for accuracy before applying glue.

Framed and carcass construction require different types of joints. Plate jointery is a new system that increases production.

Review Questions

1. What is a plain or butt joint?
2. What is a mortise-and-tenon joint?
3. What is a dovetail joint?
4. What are dowel joints? How do they compare with mortise-and-tenon joints?
5. Name the five mortise-and-tenon joints.
6. Which type of joint is more or less readily made on a power jointer?

7. The two members of a_____joint are joined at right angles, the end of one butting against the side of the other.

8. A_____joint is used mostly in making picture frames.

9. A_____joint is used to join pieces of wood end to end.

10. A_____joint is a groove cut across the grain that will receive the butt end of a piece of stock.

11. Name five of the tools used in cabinetmaking joints.

12. Why is it so important to clamp glued joints?

13. What type of glue is best for general glued joints?

14. What is a coopered joint?

15. What is a dovetail joint?

16. Why would you want to learn to make joints by hand?

17. How does plate jointery increase production?

18. The other name for an open-tenon joint is_____.

19. True or false: A depth gage for stub mortises is made by gluing a piece of paper or tape on the side of the chisel.

20. In_____construction large panels are connected with continuous joints.

21. Check all cuts for accuracy before applying_____.

Chapter 16

Mitering

By definition, a *miter* is the joint formed by two pieces of molding. Each piece is cut at an angle to match when joined angularly. Miter means to meet and match on a line bisecting the angle of junction, especially at a right angle. In other words, to cut and join the ends of two pieces obliquely at an angle is to miter.

Miter Tools

To do miter work with precision, the right tools are necessary. The first is, of course, the saw, which should be a good 20-inch backsaw of about 11 or 12 teeth to the inch, filed to a keen edge, and rubbed off on the sides with the face of an oil stone. A serviceable miter box can be made of suitable hardwood by the carpenter for most of the common miter cuts (see Figure 16-1).

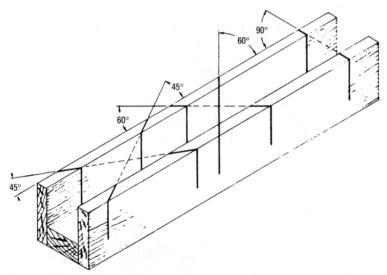

Figure 16-1 Homemade miter box.

If you want an easier job of making miters (as well as increased precision), then consider purchasing one of the many metal miter boxes available (see Figure 16-2). These come with angle settings that lock in place, saw guides, and other features that very much simplify the job. Additionally, you can obtain a motorized miter saw (see Figure 16-3). Here, all you do is lock the stock in place and

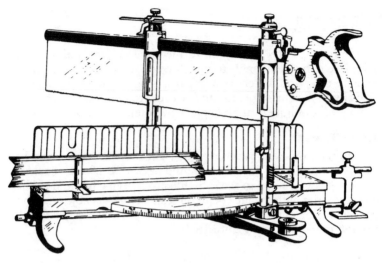

Figure 16-2   A typical metal miter box with graduated scales of angle.

Figure 16-3   Motorized miter saw. *(Courtesy of Sears, Roebuck & Co.)*

lower the blade. Of course, there are portable powered miter saws available from a number of tool manufacturers.

Moldings

In the ornamental side of carpentry construction, various forms of moldings are used. Some of these are designed to lay flush or flat against the surfaces to which they are attached (see Figure 16-4). Others are shaped to lie inclined at an angle to the nailing surfaces (see Figure 16-5). It is the *rake* or *spring* type molding that is hard to cut.

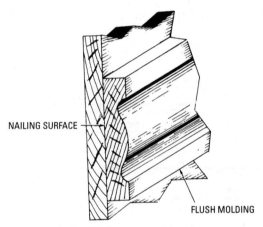

NAILING SURFACE

FLUSH MOLDING

Figure 16-4   Flush type molding.

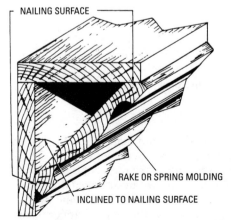

NAILING SURFACE

RAKE OR SPRING MOLDING

INCLINED TO NAILING SURFACE

Figure 16-5   Spring or rake type molding.

Mitering Flush Moldings

Where two pieces of molding join at right angles (as, for example, the sides of a picture frame), the miter angle is 45°. The term *miter angle* means the angle formed by the miter cut and edge of the molding (see Figure 16-6).

In paneling for a stairway, the moldings are joined at various angles (see Figure 16-7). This is known as *varying miters*, and a problem arises to find the miter angles. It is easily done by remembering that the miter angle is always half of the joint angle. To find the miter cut or the angle at which the miter cut is made, bisect the joint angle (see Figure 16-8). The triangle *ABC* corresponds to the triangle *A* in Figure 16-8. To find the miter cut at *A*, describe the arc *MS* of any radius with *A* as center. With *M* and *S* as centers, describe arcs *L* and *F*, intersecting at *R*. Draw a line *AR*, which is the miter cut required.

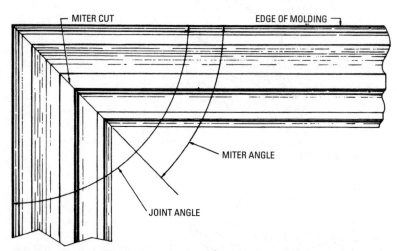

Figure 16-6 Two pieces of flush molding joined at 90°.

Mitering Spring Moldings

A *spring molding* is one that is made of thin material and is leaned or inclined away from the nailing surface (see Figure 16-5). These moldings are difficult to miter, especially when the joint is made with a gable, springs, or raking molding. The two most unusual forms of miters to cut on spring moldings are those on the inside and outside angles (see Figure 16-9). The pieces are as they would appear from the top sill looking down.

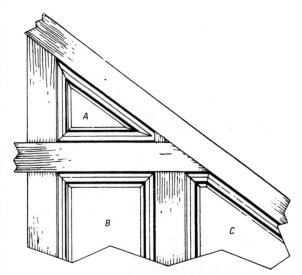

Figure 16-7 Panel work of a wall illustrating varying miters. In panel (A) each angle is different, in (B) both miters are equal, and (C) has two different angles.

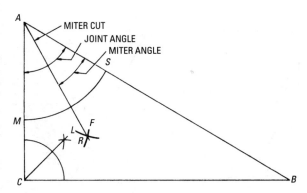

Figure 16-8 A method of finding various miter cuts for different angles.

A difficult operation for most carpenters is the cutting of a spring molding when the horizontal portion must miter with a gable or raking molding. The miter-box cuts for such joints are laid out as shown in Figure 16-10. To lay out these cuts in constructing the miter box, make the down cuts (*BB*) the same pitch as the plumb cut on the rake. The over cuts (*OO* and *O′ O′*) should be obtained

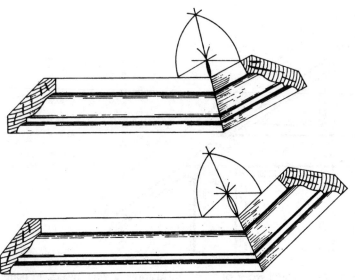

Figure 16-9 Illustration of the two most unusual miter forms to cut.

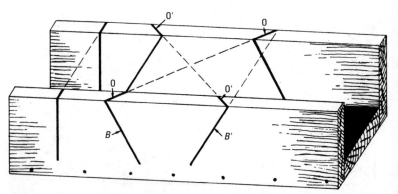

Figure 16-10 Miter box layout for cutting a spring molding when the horizontal portion has to miter with a gable or raking molding.

as follows. Suppose a roof has a ¼ pitch. Find the rafter inclination (see Figure 16-11) by laying off AB = 12-inch run and BC = 6-inch rise, giving the roof angle CAB for ¼ pitch and rafter length AC = 13.42 inch per foot run. With the setting 13.42 and 12, lay the steel square on top of the miter box (see Figure 16-12).

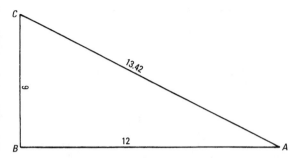

Figure 16-11 Method of finding the angle for the cuts shown in Figure 16-10.

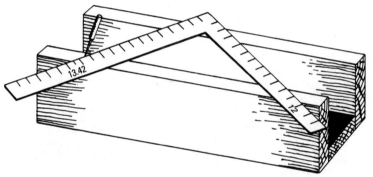

Figure 16-12 Steel square applied to the miter box with 13.42 and 12 setting to mark for cuttings.

Mitering Panel and Raised Moldings

The following instructions illustrate how raised and rabbeted moldings may be cut and inserted in panels (see Figure 16-13). *AB* denotes the outside frame, and *C* is the raised panel. *D* and *E* are the pine fillets inserted in the plowing, and *F* is the panel molding that must be mitered around the inside edges of the frame. Point *G* is the rabbet or lips on the molding *F*. If the framing *AB* is carefully constructed, and the surfaces are equal, the offset down to the panel will be equal all around, then all that is necessary is to make a hardwood strip or saddle equal in width to the depth of the offset.

The front door shown in Figure 16-14 has both flush and raised panels. A raised 1-inch molding is on the outside or street side, and an ordinary ogee and chamfer is on the inside. The enlarged

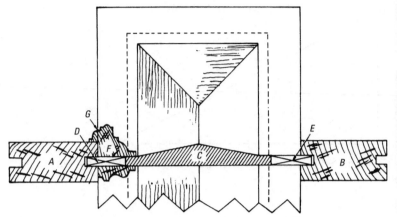

Figure 16-13 Illustrating a panel and molding design.

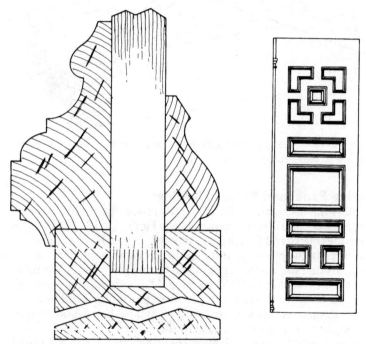

Figure 16-14 End and side view of a door with raised molding.

section is shown. This door is a good example of mitered moldings that form an attractive design. The difference between inside and outside miters must be explained. An *inside* miter is one in which the profile of the molding is contained (or rather, the outside lines and highest parts are contained) within the angle of the framing. An *outside* miter is one that is directly opposite and not contained, but the whole of the molding is mitered on the panel outside the angle. Both miters are sawed similarly in the box, with the exception of the reversing of the intersections.

Cutting Long Miters

In numerous instances, miter cuts must be made that cannot be cut in an ordinary or patent miter box. In such cases, the work is facilitated by making a special box if there are several cuts of a kind to be made.

Figure 16-15 shows a box 13 inches high and has a flare of $3\frac{1}{4}$ inches. Its construction requires miter cuts that cannot be made on an ordinary miter box. One corner is a rabbet joint, and the other

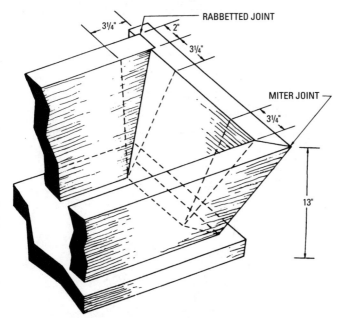

Figure 16-15 A view of two joints, one showing a rabbet joint and the other a miter joint.

corner is a miter joint. Each corner can be cut out by the use of an adjustable table power saw.

Coping

By definition, *cope* means to cover (or match against) a covering. *Coping* is generally used for moldings, the square and flat surfaces being fitted together, one piece abutting against the other. Against plaster, the inside miter is useless, since one piece is almost certain to draw away and open the joint as it is being nailed into the studding.

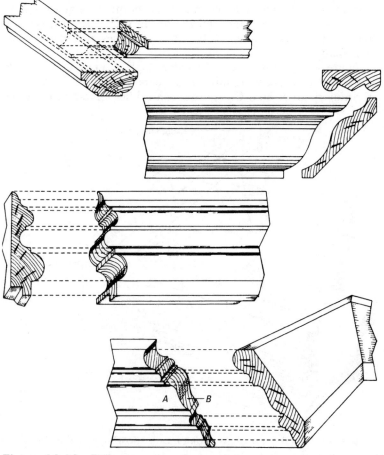

Figure 16-16 Different styles of coped joints.

It can be mitered tightly enough by cutting the lengths a little full and springing them into place, but it is not advisable except possibly in solid corners. If against plastered walls, it may crack. The best way to make this joint is to cope it.

Figure 16-16 shows various types of coped joints. To obtain this joint, the piece of molding is placed in a miter box and cut to a 45° angle. After this is done, the miter angle is cut by a coping saw along the design of the molding. If the corners are square, the miter and coped joints will fit perfectly.

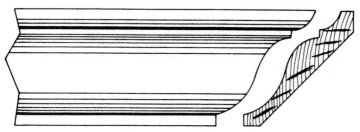

Figure 16-17　Coped crown or spring molding.

Figure 16-17 shows that, when a molding is cut in a miter box for coping, it is always the reverse of the profile, and when cut out to the line thus formed (preferably with a coping saw), it fits to it at every inside corner so as to be invisible. In brief, all curved lines and members join and intersect without interruption at any point.

Summary

To make proper miter angles, a good miter box and backsaw are needed. A carpenter can make a serviceable miter box from suitable hardwood for most of the common miter cuts.

In carpentry, various moldings are used. Some are designed to lay flat or flush against the surface, and others are shaped to lie inclined or at an angle to the surface.

Coping is another form of joining moldings. After the molding is cut at a 45° angle, a coping saw is used to saw along the design of the adjoining molding. This is done to relieve any possible strain in molding corners on walls, such as plaster.

Review Questions

1. What is a miter box?
2. What type of saw should be used in miter cuts?

3. What is coping?

4. What is spring or rake type molding?

5. Explain the difference between a rabbet joint and a miter joint.

6. What is a miter?

7. It is the _____ or spring type molding that is hard to cut.

8. An _____ miter is one in which the profile of the molding is contained (or rather, the outside lines and highest parts are contained) within the angle of the framing.

9. True or false: In many cases, miter cuts must be made that cannot be cut in an ordinary or patent miter box. They are then cut on a power table saw.

10. True or false: A steel square really has no use in the process of miter cutting.

Chapter 17

Using the Steel Square

On most construction work (especially in house framing), the steel square is invaluable for accurate measuring and for determining angles. The correct name of this tool is *framing square* because the square with its markings was designed especially for marking timber in framing. Whatever the name, the square, when used properly, is a wonderful tool.

The tool, with its various scales and tables, has been described in Chapter 3. The goal of this chapter is to explain these markings in more detail and also to explain their applications by examples showing actual uses of the square. The following names are commonly used to identify the different portions of the square:

- *Body*—The long, wide member.
- *Face*—The sides visible (both body and tongue) when the square is held by the tongue in the right hand with the body pointing to the left (see Figure 17-1).
- *Tongue*—The short, narrow member.
- *Back*—The sides visible (both body and tongue) when the

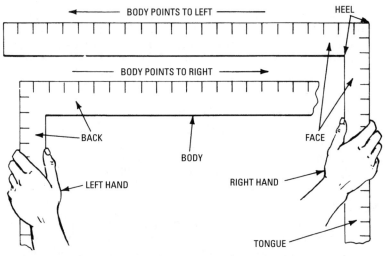

Figure 17-1 The face and back sides of a framing square. The body of the square is sometimes called the blade.

square is held by the tongue in the left hand with the body pointing to the right (see Figure 17-1).

The square most generally employed has an 18-inch tongue and a 24-inch body. The body is 2 inches wide, and the tongue is $1^{1}/_{2}$ inches wide, $^{3}/_{16}$ inch thick at the heel (or corner) for strength, diminishing (for lightness) to the two extremities to approximately $^{3}/_{32}$ inch. The various markings on squares are of two kinds:

* Scales (or graduations)
* Tables

When buying a square, it is advisable to get one with all the markings rather than a budget unit on which the manufacturer has omitted some of the scales and tables. Table 17-1 shows a comparison of the difference between an incomplete and a complete square.

Table 17-1   Square Comparisons

	Tables	*Graduations*
Cheap Square	Rafter, Essex, Brace	$^{1}/_{16}$, $^{1}/_{12}$, $^{1}/_{18}$, $^{1}/_{4}$
Complete Markings	Rafter, Essex, Brace, Octagon, Polygon cuts	$^{1}/_{100}$, $^{1}/_{64}$, $^{1}/_{32}$, $^{1}/_{16}$, $^{1}/_{12}$, $^{1}/_{10}$, $^{1}/_{8}$, $^{1}/_{4}$

The square with the complete markings will cost more, but in the purchase of tools, you should make it a rule to purchase only the finest made. The general arrangements of the markings on squares differ somewhat with different makes. It is advisable to examine the different makes before purchasing to select the one best suited to your specific requirements.

Application of the Square

As stated previously, the markings on squares of different makes sometimes vary both in their position on the square and the mode of application. However, a thorough understanding of the application of the markings on any first-class square will enable you to easily acquire proficiency with any other square.

Scale Problems

The term *scales* is used to denote the inch divisions of the tongue and body length found on the outer and inner edges. The inch graduations are divided into $^{1}/_{4}$, $^{1}/_{8}$, $^{1}/_{10}$, $^{1}/_{12}$, $^{1}/_{16}$, $^{1}/_{32}$, $^{1}/_{64}$, and $^{1}/_{100}$. All these graduations should be found on a first-class square. The various scales start from the heel of the square (that is, at the intersection of the two outer, or two inner, edges).

A square with only the scale markings is adequate to solve many problems that arise when laying out carpentry work. An idea of its range of usefulness is shown in the following problems.

Problem I
Describe a semicircle given the diameter.
Drive brads at the ends of the diameter *LF* in Figure 17-2. Place the inner edges of the square against the nails, and hold a lead pencil at the inner heels *S*. Then rotate the square, sliding the inner edges along the brads, marking the semicircle with the pencil.

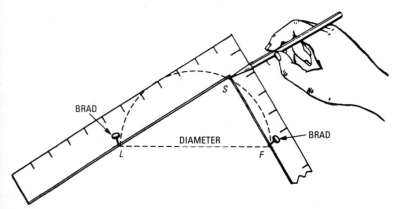

Figure 17-2   Problem 1. Describing a semicircle. Rotate the square, sliding the inner edges along the brads, marking a semicircle with the pencil.

Problem 2
Find the center of a circle.
Lay the square on the circle so that its outer heel lies in the circumference. Mark the intersections of the body and tongue with the circumference. The line that connects these two points is a *diameter*. Draw another diameter (obtained in the same way). The intersection of the two diameters is the center of the circle (see Figure 17-3).

Problem 3
Describe a circle through three points that are not in a straight line.
Join the three points with straight lines. Bisect these lines, and, at the points of bisection, erect perpendiculars with the square. The intersection of these perpendiculars is the center from which a circle may be described through the three points (see Figure 17-4).

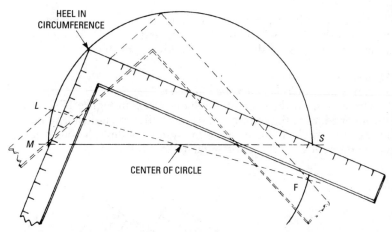

Figure 17-3 Problem 2. Draw diameters through points LF and MS, where the sides of the square touch the circle with the heel in the circumference. The intersection of these two lines is the center of the circle.

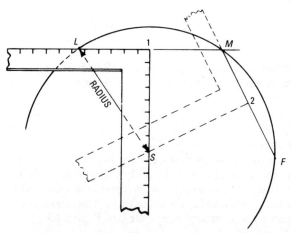

Figure 17-4 Problem 3. Let points L, M, and F be three points that are not in a straight line. Draw lines LM and MF, and bisect them at points 1 and 2, respectively. Apply the square with the heel at points 1 and 2 as shown. The intersection of the perpendicular lines thus obtained, point S is the center of the circle. Lines LS, MS, and FS represent the radius of the circle, which may now be described through points L, M, and F.

Problem 4
Find the diameter of a circle whose area is equal to the sum of the areas of two given circles.

Lay off on the tongue of the square the diameter of one of the given circles and on the body the diameter of the other circle. The distance between these points (measure across with a 2-foot rule) will be the diameter of the required circle (see Figure 17-5).

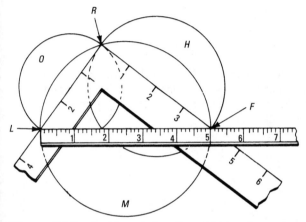

Figure 17-5   Problem 4. Let O and H be the two given circles, with their diameters LR and RF at right angles. Suppose the diameter of O is 3 inches and the diameter of H is 4 inches. Points L and F, at these distances from the heel of the square, will be 5 inches apart, as measured with a 2-foot rule. This distance LF, or 5 inches, is the diameter of the required circle. Proof: $(LF)^2 = (LR)^2 + (RF)^2$, or $25 = 9 + 16$.

Problem 5
Lay off angles of 30° and 60°.

Mark off 15 inches on a straight line, and lay the square so that the body touches one end of the line and the $7\frac{1}{2}$-inch mark on the tongue is against the other end of the line (see Figure 17-6). The tongue will then form an angle of 60° with the line, and the body will form an angle of 30° with the line.

Problem 6
Lay off an angle of 45°.

The diagonal line connecting equal measurements on either arm of the square forms angles of 45° with the blade and tongue (see Figure 17-7).

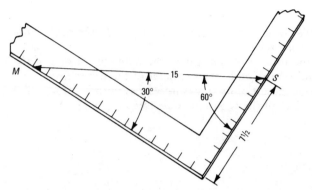

Figure 17-6 Problem 5. Draw line MS, 15 inches long. Place the square so that point S touches the tongue 7½ inches from the heel and point M touches the body. The triangle thus formed will have an angle of 30° at M and an angle of 60° at S.

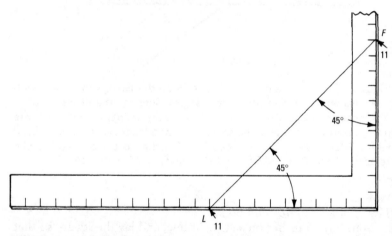

Figure 17-7 Problem 6. Take equal measurements L and F on the body and tongue of the square. The triangle thus formed will have an angle of 45° at L and at F.

Problem 7
Lay off any angle.

Table 17-2 gives the values for measurements on the tongue and the body of the square so that by joining the points corresponding to the measurements, any angle may be laid out from 1° to 45° (see Figure 17-8).

Table 17-2   Angle Table for the Square

Angle	Tongue	Body	Angle	Tongue	Body	Angle	Tongue	Body
1	0.35	20.00	16	5.51	19.23	31	10.28	17.14
2	0.70	19.99	17	5.85	19.13	32	10.60	16.96
3	1.05	19.97	18	6.18	19.02	33	10.89	16.77
4	1.40	19.95	19	6.51	18.91	34	11.18	16.58
5	1.74	19.92	20	6.84	18.79	35	11.47	16.38
6	2.09	19.89	21	7.17	18.67	36	11.76	16.18
7	2.44	19.85	22	7.49	18.54	37	12.04	15.98
8	2.78	19.81	23	7.80	18.40	38	12.31	15.76
9	3.13	19.75	24	8.13	18.27	39	12.59	15.54
10	3.47	19.70	25	8.45	18.13	40	12.87	15.32
11	3.82	19.63	26	8.77	17.98	41	13.12	15.09
12	4.16	19.56	27	9.08	17.82	42	13.38	14.89
13	4.50	19.49	28	9.39	17.66	43	13.64	14.63
14	4.84	19.41	29	9.70	17.49	44	13.89	14.39
15	5.18	19.32	30	10.00	17.32	45	14.14	14.14

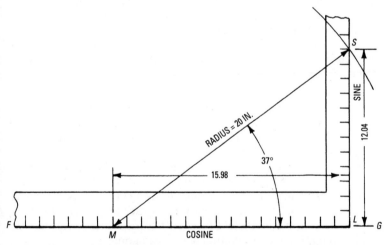

Figure 17-8   Problem 7. Let 37° be the required angle. Place the body of the square on line FG, and, from Table 17-1, lay off LS (12.04) on the tongue and LM (15.98) on the body. Draw line MS; then angle LMS = 37°. Line MS will be found to be equal to 20 inches for any angle because the values given in Table 17-1 for LS and MS are natural sines and cosines multiplied by 20.

Problem 8

Find the octagon of any size of timber.

Place the body of a 24-inch square diagonally across the timber so that both extremities (ends) of the body touch opposite edges. Make a mark at 7 inches and 17 inches, as shown in Figure 17-9. Repeat this process at the other end, and draw lines through the pairs of marks. These lines show the portion of material that must be taken off the corners.

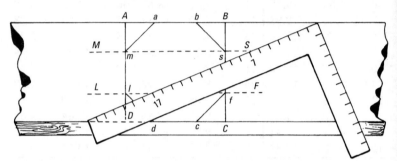

Figure 17-9 Problem 8. Lay out square ABCD. Place the body of a 24-inch square as shown, and draw parallel lines MS and LF through points 7 and 17. These lines intercept sides ml and sf of the octagon. Lay off side sb, place the square so that the tongue touches point S and the body touches I, with the heel touching line AB. The remaining sides are obtained in a similar manner.

The side of an inscribed octagon can be obtained from the side of a given square by multiplying the side of the square by 5 and dividing the product by 12. The quotient will be the side of the octagon. Figure 17-10 illustrates this method.

The side of a hexagon is equal to the radius of the circumscribing circle. If the side of a desired hexagon is given, arcs should be struck from each extremity at a radius equal to its length. The point where these arcs intersect is the center of the circumscribing circle, and having described it, it is sufficient to lay off cords on its circumference equal to the given side to complete the hexagon.

Square-and-Bevel Problems

By the application of a large bevel to the framing square, the combined tool becomes a calculating machine, and by its use, arithmetical processes are greatly simplified. The bevel is preferably made of steel blades.

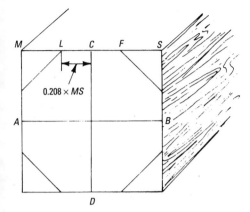

Figure 17-10 Problem 8 (second method). Let lines AB and CD be centerlines, and let line MS be one side of the square timber. Multiply the length of the side by 0.208. The product is half the side of the inscribed octagon. Therefore, lay off CF and CL, each 0.208 times side MS. LF is then one side of the octagon. Set dividers to distance CL, and lay off the other sides of the octagon from the centerlines to complete the octagon.

Note

The edges of each blade must be true. Blade E in Figure 17-11 must lie under the square so that it does not hide the graduations. The two blades must be fastened by a thumbscrew to lock them together. Blade L should have a hole near each end and one in the middle, so that blade E may be shifted as required, with a large notch near each hole in order to observe the position of blade E.

Problem 9

Find the diagonal of a square.

Set blade E to $10^3/_8$ on the tongue and to 15 on the body. Assume an 8-inch square. Slide the bevel sideways along the tongue until blade E is against point 8. The other edge will touch $11^5/_{16}$ on the body. This is the required diagonal.

Problem 10

Find the circumference of a circle from its diameter.

Set the bevel blade to 7 on the tongue of the square and to 22 on the body. The reading on the body will be the circumference corresponding to the diameter at which E is set on the tongue. To reverse the process, use the same bevel, and read the required diameter from the tongue, the circumference being set on the body.

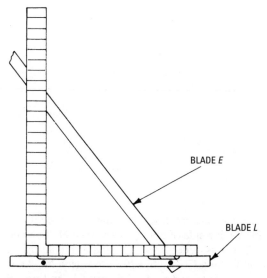

Figure 17-11 The application of a bevel to the square for solving square-and-bevel problems.

Problem 11
Given the diameter of a circle, find the side of a square of equal area.

Set the bevel blade to $10\frac{5}{8}$ on the tongue and to 12 on the body. The diameter of the circle, on the body, will give the side of the equal square on the tongue. If the circumference is given instead of the diameter, set the bevel to $5\frac{1}{2}$ on the tongue and to $19\frac{1}{2}$ on the body, thereby finding the side of the square on the tongue.

Problem 12
Given the side of a square, find the diameter of a circle of equal area.

Using the same bevel as in Problem 11, blade E is set to the given side on the tongue of the square, and the required diameter is read off the body.

Problem 13
Given the diameter of the pitch circle of a gear wheel and the number of teeth, find the pitch.

Take the number of teeth (or a proportional part) on the body of the square and the diameter (or a similar proportional part) on the tongue, and set the bevel blade to those marks. Slide the bevel to 3.14 on the body, and the number given on the tongue multiplied by the proportional divisor will be the required pitch.

Problem 14
Given the pitch of the teeth and the diameter of the pitch circle in a gear wheel, find the number of teeth.

Set the bevel blade to the pitch on the tongue and to 3.14 on the body of the square. Move the bevel until it marks the diameter on the tongue. The number of teeth can then be read from the blade. If the diameter is too large for the tongue, divide it and the pitch into proportional parts, and multiply the number found by the same figure.

Problem 15
Given the side of a polygon, find the radius of the circumscribing circle.

Set the bevel to the pairs of numbers in Table 17-3, taking $\frac{1}{8}$ or $\frac{1}{10}$ of an inch as a unit. The bevel, when locked, is slid to the given length of the side, and the required length of the radius is read on the other leg of the square. For example, if a pentagon (5 sides) must be laid out with a side of 6 inches, the bevel is set to the figures in column 5 with the lesser number set on the tongue. In this case, $\frac{74}{8} = 9\frac{1}{4}$ on the tongue, and $\frac{87}{8} = 10\frac{7}{8}$ on the body of the square. Slide the bevel to 6 on the body. The length of the radius, $5\frac{3}{32}$, will be read on the tongue.

Table 17-3   Inscribed Polygons

Number of Sides	3	4	5	6	7	8	9	10	11	12
Radius	56	70	74	60	60	98	22	89	80	85
Side	97	99	87	60	52	75	15	95	45	44

Problem 16
Divide the circumference of a circle into a given number of equal parts.

From the column marked Y in Table 17-4, take the number opposite the given number of parts. Multiply this number by the radius of the circle. The product will be the length of the cord to lay off on the circumference.

Problem 17
Given the length of a cord, find the radius of the circle.

This is the same as Problem 16, but the present form may be more expeditious for calculations. The method is useful for determining the diameter of gear wheels when the pitch and number of teeth

Table 17-4 Cords or Equal Parts

No. of Parts		Y	Z
3	Triangle	1.732	0.5773
4	Square	1.414	0.7071
5	Pentagon	1.175	0.8006
6	Hexagon	1.000	1.0000
7	Heptagon	0.8677	1.1520
8	Octagon	0.7653	1.3065
9	Nonagon	0.6840	1.4619
10	Decagon	0.6180	1.6184
11	Undecagon	0.5634	1.7747
12	Duodecagon	0.5176	1.9319
13	Tridecagon	0.4782	2.0911
14	Tetradecagon	0.4451	2.2242
15		0.4158	2.4050
16		0.3902	2.5628
17		0.3675	2.7210
18		0.3473	2.8793
19		0.3292	3.0376
20		0.3129	3.1962
22		0.2846	3.5137
24		0.2610	3.8307
25		0.2506	3.9904
27		0.2322	4.3066
30		0.2090	4.7834
36		0.1743	5.7368
40		0.1569	6.3728
45		0.1395	7.1678
50		0.1256	7.9618
54		0.1163	8.5984
60		0.1047	9.5530
72		0.0872	11.462
80		0.0785	12.733
90		0.0698	14.327
100		0.0628	15.923
108		0.0582	17.182
120		0.0523	19.101
150		0.0419	23.866

have been given. Multiply the length of the cord, width of the side, or pitch of the tooth by the figures found corresponding to the number of parts in column Z of Table 17-4. The result is the radius of the desired circle.

Table Problems

The term *table* is used here to denote the various markings on the framing square with the exception of the scales already described. Since these tables relate mostly to problems encountered in cutting lumber for roof-frame work, it is first necessary to know something about roof construction to become familiar with the names of the various rafters and other parts. Figure 17-12 is a view of a roof frame showing the various members. Note that there is a plate at the bottom and a ridge timber at the top. These are the main members to which the rafters are fastened.

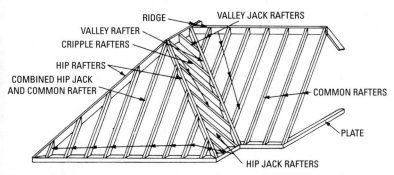

Figure 17-12 A typical roof frame, showing the ridge, the plate, and various types of rafters.

Main or Common Rafters

The following definitions relating to rafters should be carefully noted:

- The *rise* of a roof is the distance found by following a plumb line from a point on the central line of the top of the ridge to the level of the top of the plate.

- The *run* of a common rafter is the shortest horizontal distance from a plumb line through the center of the ridge to the outer edge of the plate.

- The *rise per foot run* is the basis on which rafter tables on some squares are made. The term is self-defining. Figure 17-13 shows other roof components.

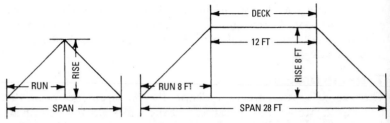

Figure 17-13 The terms rise, run, span, and deck are illustrated in two types of roofs. If the rafters rise to a deck instead of a ridge, subtract the width of the deck from the span. For example, assume the span is 28 feet and the deck is 12 feet. The difference is 16 feet, and the pitch = 8/(28-12) = $\frac{1}{2}$.

To obtain the rise per foot run, multiply the rise by 12 and divide by the run, as follows:

$$\text{rise per foot run} = \frac{\text{rise} \times 12}{\text{run}}$$

The factor 12 is used to obtain a value in inches, since the rise and run are normally given in feet.

Problem 18
If the rise is 8 feet and the run is 8 feet, what is the rise per foot run?

$$\text{rise per foot run} = \frac{8 \times 12}{8} = 12 \text{ inches}$$

The rise per foot run is always the same for a given pitch and can be remembered for all ordinary pitches, as shown in Table 17-5.

Table 17-5 Rise per Foot Run

Pitch	$\frac{1}{2}$	$\frac{1}{3}$	$\frac{1}{4}$	$\frac{1}{6}$
Rise per foot run (inches)	12	8	6	4

The pitch can be obtained if the rise and run are known (Figure 17-14) by dividing the rise by twice the run as follows:

$$\text{pitch} = \frac{\text{rise}}{2 \times \text{run}}$$

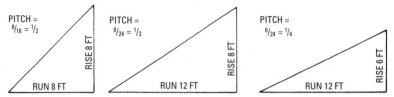

Figure 17-14 To obtain the pitch of any roof divide the rise of the rafters by twice the run.

In roof construction, the rafter ends are cut with slants that rest against the ridge and the plate (see Figure 17-15A). The slanting cut that rests against the ridge board is called the *plumb* (or *top*) cut, and the cut that rests on the plate is called the *seat* (or *heel*) cut.

The length of the common rafter is the length of a line from the outer edge of the plate to the top corner of the ridge board or, if there is no ridge board, from the outer edge of the plate to the vertical centerline of the building (see Figure 17-15B). The run of the rafter, then, in the first case is half the width of the building less half the thickness of the ridge, if any. If there is no ridge board, the run is half the width of the building. Where there is a deck, the run of the rafters is half the width of the building less half the width of the deck.

Now, with a 24-inch square, draw diagonals connecting 12 on the tongue (corresponding to the run) to the value from Table 17-6 on the body (corresponding to the rise) to obtain the pitch angle for any combination of run and rise. Figure 17-16 further illustrates this procedure.

Hip Rafters

The hip rafter represents the hypotenuse (or diagonal) of a right-angle triangle. One side is the common rafter, and the other side is the plate, or that part of the plate lying between the foot of the hip rafter and the foot of the adjoining common rafter (see Figure 17-17).

The rise of the hip rafter is the same as that of the common rafter. The run of the hip rafter is the horizontal distance from the plumb line of its rise to the outside of the plate at the foot of the hip rafter. This run of the hip rafter is to the run of the common rafter as 17 is to 12. Therefore, for a $\frac{1}{6}$ pitch, the common rafter run and rise are 12 and 4, respectively, whereas the hip rafter run and rise are 17 and 4, respectively.

For the top and bottom cuts of the common rafter, the figures are used that represent the common rafter run and rise (that is, 12 and

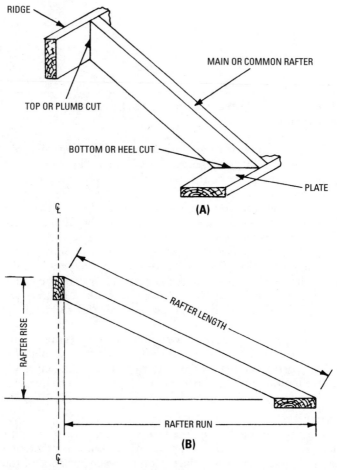

Figure 17-15 A portion of the roof frame, showing the top (or plumb) cut and the bottom (or heel) cut is illustrated in A. The length of a common rafter is shown in B.

Table 17-6 Pitch Table

Pitch	I	$11/_{12}$	$5/_6$	$3/_4$	$2/_3$	$7/_{12}$	$1/_2$	$5/_{12}$	$1/_3$	$1/_4$	$1/_6$	$1/_{12}$
Run	12	12	12	12	12	12	12	12	12	12	12	12
Rise	24	22	20	18	16	14	12	10	8	6	4	2

4 for a $\frac{1}{6}$ pitch, 12 and 6 for a $\frac{1}{4}$ pitch, and so on). However, for the top and bottom cuts of the hip rafter, use the figures 17 and 4, 17 and 6, and so on, as the run and rise of the hip rafter. It must be remembered, however, that these figures will not be correct if the pitches on the two insides of the hip (or valley) are not the same.

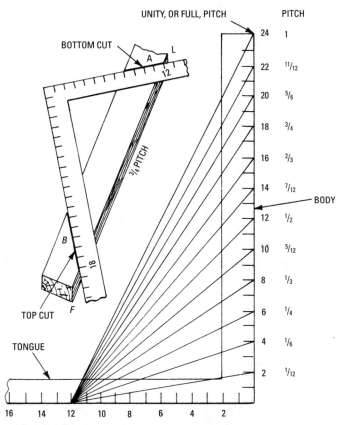

Figure 17-16   The application of the framing square for obtaining the various pitches given in Table 17-6.

Valley Rafters

The valley rafter is the hypotenuse of a right-angle triangle made by the common rafter with the ridge, as shown in Figure 17-18. This corresponds to the right-angle triangle made by the hip rafter with the common rafter and plate. Therefore, the rules for the lengths and cuts of valley rafters are the same as for hip rafters.

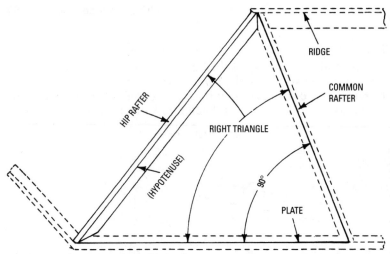

Figure 17-17 The hip rafter is framed between the plate and the ridge and is the hypotenuse of a right-angle triangle whose other two sides are the adjacent common rafter and the intercepted portion of the plate.

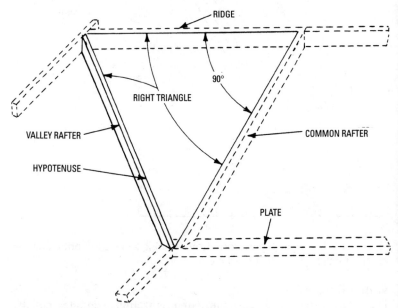

Figure 17-18 The valley rafter is framed between the plate and the ridge and is the hypotenuse of a right-angle triangle whose other two sides are the adjacent common rafter and the intercepted portion of the ridge board.

Jack Rafters

These are usually spaced either 16 or 24 inches apart, and, since they lie equally spaced against the hip or valley, the second jack rafter must be twice as long as the first, the third three times as long as the first, and so on (see Figure 17-19). One reason for the 16- and 24-inch spacings on jack rafters is because of the roof sheathing. Therefore, the rafters must be 16 or 24 inches apart so that the sheathing may be conveniently nailed to them.

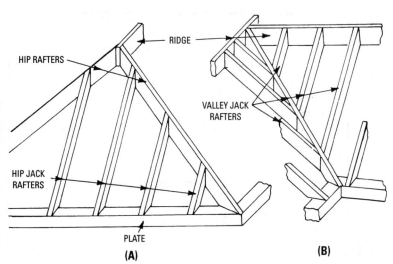

Figure 17-19 Jack rafters. (A) Hip jack rafters, framed between the plate and hip rafters and (B) valley jack rafters, framed between the ridge and the valley rafter.

Cripple Rafters

A cripple rafter is a jack rafter that touches neither the plate nor the ridge. It extends from the valley rafter to the hip rafters. The cripple-rafter length is that of the jack rafter plus the length necessary for its bottom cut, which is a plumb cut similar to the top cut. Top and bottom (plumb) cuts of cripples are the same as the top cut for jack rafters. The side cut at the hip and valley is the same as the side cut for jacks.

Finding Rafter Lengths Without the Aid of Tables

In the directions accompanying some framing squares and in some books, frequent mention is made of the figures 12, 13, and 17. Directions say, for example, that for common rafters "use figure 12 on the body and the rise of the roof on the tongue"; for hip or valley rafters, "use figure 17 on the body and the rise of the roof on the tongue"—but no explanation of how these fixed numbers are obtained is provided.

The origin of such fixed numbers should be known to make the typical job easier to understand. They can be readily understood by referring to Figure 17-20. In this illustration, let ABCD be a square

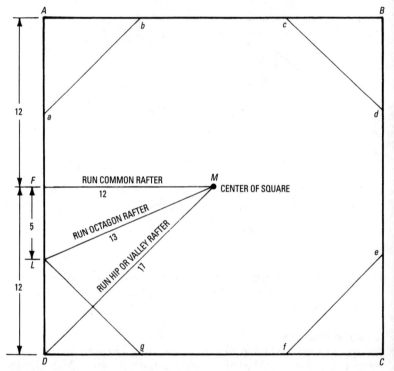

Figure 17-20 A square and an inscribed octagon are used to illustrate the method of obtaining and using points 12, 13, and 17 in the application of a framing square to determine the length of rafters without the aid of rafter tables.

whose sides are 24 inches long, and let abcdefgL be an inscribed octagon. Each side of the octagon (ab, bc, and so on) measures 10 inches (that is, LF = one-half side = 5 inches, and by construction, FM = 12 inches). Now, let FM represent the run of a common rafter. Then LM will be the run of an octagon rafter, and DM will be the run of a hip or valley rafter. The values for the run of octagon and hip or valley rafters (LM and DM, respectively) are obtained as follows:

$$LM = \sqrt{(FM)^2 + (LF)^2} = \sqrt{(12)^2 + (5)^2} = 13$$

$$DM = \sqrt{(FM)^2 + (DF)^2} = \sqrt{(12)^2 + (12)^2} = 16.97$$

(or approximately 17)

Problem 19
What is the length of a common rafter having a 10-foot run and a ³/₈ pitch?
For a 10-foot run,

the span $= 2 \times 10 = 20$ feet

with ³/₈ pitch,

rise $= $ ³/₈ $\times 20 = 7.5$ feet

rise per foot run $= \dfrac{\text{rise} \times 12}{\text{run}} = \dfrac{7.5 \times 12}{10} = 9$ inches

On the body of the square (see Figure 17-21), take 12 inches for 1 foot of run, and on the tongue, take 9 inches for the rise per foot of run. The diagonal (or distance between the points thus obtained) will be the length of the common rafter per foot of run with a ³/₈ pitch. The distance FM measures 15 inches, or by calculation:

$$FM = \sqrt{(12)^2 + (9)^2} = 15 \text{ inches}$$

Since the length of run is 10 feet, the following is true:

length of rafter $=$ length of run $\times$ length per foot

$$= 10 \times {}^{15}/_{12}$$

$$= \frac{150}{12}$$

$$= 12.5 \text{ feet}$$

The combination of figures 12 and 9 on the square (see Figure 17-21) not only gives the length of the rafter per foot of run but

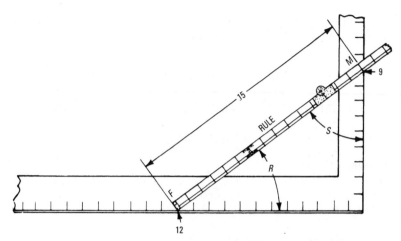

Figure 17-21 A rule is placed on the square at points 12 and 9 to obtain the length of a common rafter per foot of run with a $\frac{3}{8}$ pitch.

also, if the rule is considered as the rafter, the angles S and R for the top and bottom cuts are obtained. The points for making the top and bottom cuts are found by placing the square on the rafter so that a portion of one arm of the square represents the run and a portion of the other arm represents the rise. For the common rafter with a $\frac{3}{8}$ pitch, these points are 12 and 9. The square is placed on the rafter as shown in Figure 17-22.

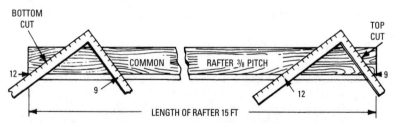

Figure 17-22 The square is placed on the rafter at points 12 and 9, as shown, thereby giving the proper angles for the bottom and top cuts.

Problem 20
What length must an octagon rafter be to join a common rafter having a 10-foot run (as rafters MF and ML in Figure 17-20)?

From Figure 17-20, it is seen that the run per foot of an octagon rafter, as compared with a common rafter, is as 13 is to 12 and that

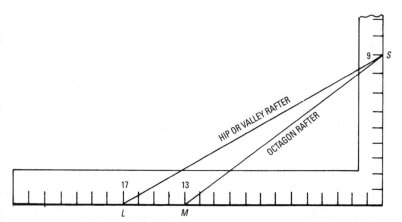

Figure 17-23 Measurements using the square for octagon and hip or valley rafters, illustrating the use of points 13 and 17. Line MS (13,9) is the octagon rafter length per foot of run of a common rafter with a $3/8$ pitch; line LS (17,9) is the hip or valley rafter length per foot of run of a common rafter with a $3/8$ pitch.

the rise for a 13-inch run of an octagon rafter is the same as for the run of a 12-inch common rafter. Therefore, measure across from points 13 and 9 on the square, as MS in Figure 17-23, which gives the length ($15^3/4$ inches) of an octagon rafter per foot of run of a common rafter. The length multiplied by the run of a common rafter gives the length of an octagon rafter, as follows:

$$15^3/4 \times 10 = 157^1/2 \text{ inches} = 13 \text{ feet, } 1^1/2 \text{ inches}$$

Points 13 and 9 on the square (MS in Figure 17-23) give the angles for the top and bottom cuts.

Problem 21
What length must a hip or valley rafter be to join a common rafter having a 10-foot run (as rafters MS and MD in Figure 17-20)?

Figure 17-20 shows that the run per foot of a hip or valley rafter, as compared with a common rafter, is as 17 is to 12 and that the rise per 17-inch run of a hip or valley rafter is the same as for a 12-inch run of a common rafter. Therefore, measure across from points 17 and 9 on the square, as LS in Figure 17-23. This gives the length ($19^1/4$ inches) of the hip or valley rafter per foot of common rafter. This length, multiplied by the run of a common rafter, gives the length of the hip or valley rafter.

$19^{1}/_{4} \times 10 = 19^{1}/_{2}$ inches $= 16$ feet, $^{1}/_{2}$ inch

Points 17 and 9 on the square (LS in Figure 17-23) give the angles for the top and bottom cuts.

Table 17-7 gives the points on the square of the top and bottom cuts of various rafters.

Table 17-7   Square Points for Top and Bottom Cuts

Pitch	I	$^{11}/_{12}$	$^{5}/_{6}$	$^{3}/_{4}$	$^{2}/_{3}$	$^{7}/_{12}$	$^{1}/_{2}$	$^{5}/_{12}$	$^{1}/_{3}$	$^{1}/_{4}$	$^{1}/_{6}$	$^{1}/_{12}$
Tongue:												
Common							12					
Octagon							13					
Hip or Valley							17					
Body	24	22	20	18	16	14	12	10	8	6	4	2

Rafter Tables

The arrangement of these tables varies considerably with different makes of squares, not only in the way they are calculated but also in their positions on the square. On some squares, the rafter tables are found on the face of the body; on others, they are found on the back of the body. There are two general classes of rafter tables, grouped as follows:

* Length of rafter per foot of run
* Total length of rafter

Evidently, where the total length is given, there is no figuring to be done, but when the length is given per foot of run, the reading must be multiplied by the length of run to obtain the total length of the rafter. To illustrate these differences, directions for using several types of squares are given in the following sections. These differences relate to the common and hip or valley rafter tables.

Reading the Total Length of the Rafter

One popular type of square is selected as an example to show how rafter lengths may be read directly without any figuring. The rafter tables on this particular square occupy both sides of the body instead of being combined in one table; the common rafter table is found on the back, and the hip, valley, and jack rafter tables are located on the face.

Common Rafter Table

The common rafter table, Figure 17-24, includes the outside-edge graduations of the back of the square on both the body and the tongue. These graduations are in twelfths. The inch marks may represent inches or feet, and the twelfths marks may represent $1/12$ of an inch or $1/12$ of a foot (inches). The edge-graduation figures above the table represent the run of the rafter; under the proper figure on the line representing the pitch is found the rafter length required in the table. The pitch is represented by the figures at the left of the table under the word *PITCH*. Table 17-8 shows the pitch for 12 feet of run.

Table 17-8   Feet of Run

Feet of Rise	4	6	8	10	12	15	18
Pitch	$1/6$	$1/4$	$1/3$	$5/12$	$1/2$	$5/8$	$3/4$

The length of a common rafter given in the common rafter table is from the top center of the ridge board to the outer edge of the plate. In actual practice, deduct half the thickness of the ridge board, and add for any eave projection beyond the plate.

Problem 22

Find the length of a common rafter for a roof with a $1/6$ pitch (rise = $1/6$ the width of the building) and a run of 12 feet (found in the common rafter table, Figure 17-24, the upper, or $1/6$-pitch ruling).

Find the rafter length required under the graduation figure 12. This is found to be 12, 7, 10, which means 12 feet, $7^{10}/12$ inches. If the run is 11 feet and the pitch is $1/2$ (the rise = $1/2$ the width of the building), then the rafter length will be 15, 6, 8, which means 15 feet $6^{8}/12$ inches. If the run is 25 feet, add the rafter length for a run of 20 feet to the rafter length for a run of 5 feet. When the run is in inches, then in the rafter table read inches and twelfths instead of feet and inches. For example, if, with a $1/2$ pitch, the run is 12 feet 4 inches, add the rafter length of 4 inches to that of 12 feet as follows:

For a run of 12 feet, the rafter length is 16 feet, $11^{8}/12$ inches

For a run of 4 inches, the rafter length is $5^{8}/12$ inches

The total is 17 feet, $5^{8}/12$ inches

The run of 4 inches is found under the graduation 4 and is 5, 7, 11, which is approximately $5^{8}/12$ inches. If the run were 4 feet, it would be read as 5 feet, $7^{11}/12$ inches.

Hip Rafter Table

This table (see Figure 17-25) is located on the face of the body and is used in the same manner as the table for common rafters explained earlier. In the hip rafter table, the outside-edge-graduation figures represent the run of common rafters. The length of a rafter given in the table is from the top center of the ridge board to the outer edge of the plate. In actual practice, deduct half the thickness of the ridge board, and add for any eave projection beyond the plate. When using this table, find the figures on the line with the required pitch of the roof.

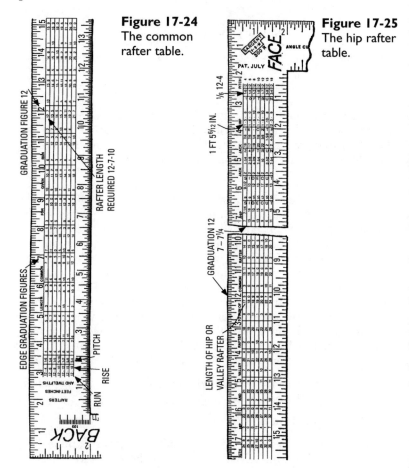

Figure 17-24
The common rafter table.

Figure 17-25
The hip rafter table.

Under *PITCH*, the set of three columns of figures gives the pitch. The seven pitches in common use are given. For example, $\frac{1}{6}$-12-4 means that for a $\frac{1}{6}$ pitch, there is a 12-inch run per 4-inch rise.

Under *HIP*, the set of figures gives the length of the hip and valley rafter per foot of run of common rafter for each pitch, such as 1 foot, $5^{6/12}$ inches for a $\frac{1}{6}$ pitch.

Under *JACK* (16 inches on center), the set of figures gives the length of the shortest jack rafter, spaced 16 inches on center, which is also the difference in length of succeeding jack rafters.

Problem 23
If the jack rafters are spaced 16 inches on center for a $\frac{1}{6}$-pitch roof, find the lengths of the jacks and cut bevels.

The jack top and bottom cuts (or plumb and heel cuts) are the same as for the common rafter. Take 12 on the tongue of the square (that is, mark on the $9\frac{1}{2}$ sides, as shown in the illustration), which represents the rise per foot of the roof. If the pitch is given, take the figures in Table 17-7 that correspond to the given pitch. Thus, for a $\frac{1}{6}$ pitch, these points are 12 and 4. Figure 17-26 shows the square on the jack in this position for marking top and bottom cuts.

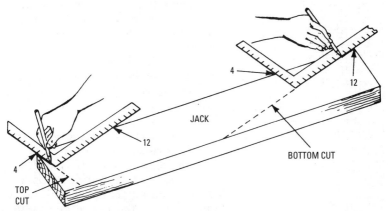

Figure 17-26   The square is applied to a jack rafter for marking top and bottom cuts. The vertical and horizontal cuts for jack rafters are the same as for common rafters.

Look along the line of $\frac{1}{6}$ pitch, in Figure 17-27, under *JACK* (16-inch center), and find $16^{7}/_{8}$, which is the length in inches of the shortest jack and is also the amount to be added for the second jack.

Deduct half the thickness of the hip rafter because the jack rafter
lengths given in this table are to centers. Also, add for any projection
beyond the outer edge of the plate.

Look along the line of $^1/_6$ pitch, in Figure 17-27, under *JACK*
(side cut), and find 9–9$^1/_2$ for a $^1/_6$ pitch. These figures refer to the
graduated scale on the edge of the

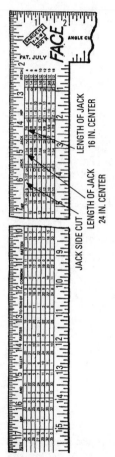

square. To obtain the required bevel,
take 9 on one arm and 9$^1/_2$ on
the other, as shown in Figure 17-28.
It should be carefully noted that the
last figure (or figure to the right) gives
the point on the marking side of the
square (that is, mark on the 9$^1/_2$ sides,
as shown in the illustration).

Under *JACK* (24 inches on center),
the set of figures gives the length of the
shortest jack rafter spaced 24 inches
on center, which is also the difference
in length of succeeding jack rafters.
Deduct half the thickness of the hip
or valley rafter because the jack rafter
lengths given in the table are to centers. Also, add for any projection beyond the plate.

Under *HIP,* the set of figures gives
the side cut of the hip and valley rafters
against the ridge board or deck, as
7–7$^1/_4$ for a $^1/_6$ pitch (mark on the
7$^1/_4$ side).

To get the cut of the sheathing and
shingles (whether hip or valley), reverse the figures under *HIP,* as 7$^1/_4$–7
instead of 7–7$^1/_4$. For the hip top and
bottom cuts, take 17 on the body of the
square, and, on the tongue, take the figure that represents the rise per foot of
the roof.

Figure 17-27 A rafter
table.

Figure 17-29 shows the marking and
cut of the hip rafter, and Figure 17-30
shows the rafter in position resting on the cap and the ridge. The
section LARF resting on ridge is the same as LARF (Figure 17-29).

Under *HIP AND VALLEY,* the set of figures gives the length
of run of the hip or valley rafter for each pitch of the common

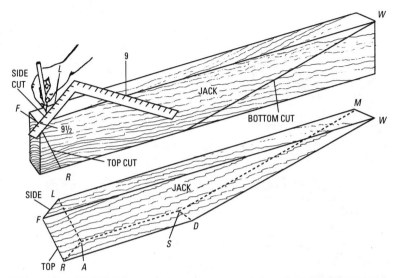

Figure 17-28 Marking and cutting a jack rafter with the aid of the square. FR and DW are the marks for the top and bottom cuts, respectively. With the jack rafter cut as marked, LARF represents the section cut at the top, and MSDW represents the section cut at the bottom.

rafter. For instance, for a roof with a $1/6$ pitch under the figure 12 (representing the run of the common rafter, or half the width of the building), along the $1/6$-pitch line of figures find 17, 5, 3, which means 17 feet, $5^3/_{12}$ inches, which is the length of the hip or valley rafter. Deduct half the thickness of the ridge board, and add for eave overhang beyond the plate, which is the length of the hip or valley rafter required for a roof with a $1/6$ pitch and a common rafter run of 12 feet.

Problem 24

Find the length of the hip rafter for a building that has a 24-foot span and a $1/6$ pitch (a 4-inch rise per foot of run).

In the hip rafter table (see Figure 17-25) along the line of figures for $1/6$ pitch and under the graduation figure 12 (representing half the span, or the run of the common rafter), find 17, 5, 3, which means 17 feet, $5^3/_{12}$ inches. This is the required length of the hip or valley rafter. Deduct half the thickness of the ridge board, and add for any overhang required beyond the plate.

For the top and bottom cuts of the hip or valley rafter, take 17 on the body of the square and 4 (the rise of the roof per foot) on the

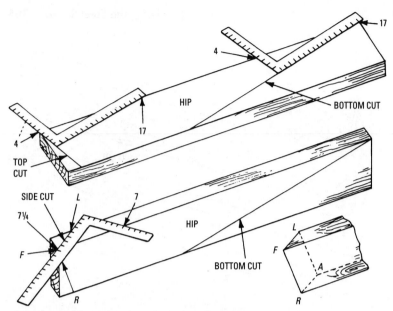

Figure 17-29 The square, as applied to hip rafters, for marking top, bottom, and side cuts. Note that the number 17 on the body is used for hip rafters. Section LARF shows the bevel required for the ridge.

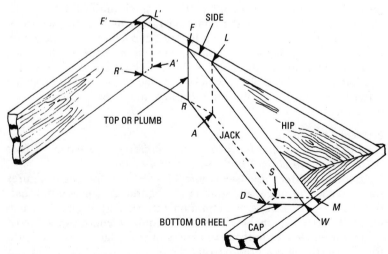

Figure 17-30 The jack rafter in position on the roof between the hip rafter and the cap.

tongue. The mark on the 17 side gives the bottom cut. The mark on the 4 side gives the top cut.

For the side cut of the hip or valley rafter against the ridge board, look in the set of figures for the side cut in the table (see Figure 17-25) under *HIP* along the line for $1/6$ pitch, and find the figure 7–7$1/4$. Use 7 on one arm of the square and 7$1/4$ on the other. Mark on the 7$1/4$ arm for the side cut.

Reading Length of Rafter per Foot of Run

There are many methods used by carpenters for determining the lengths of rafters, but probably the most dependably accurate method is the length-per-foot-of-run method. Since many (perhaps most) of the better rafter-framing squares now have tables on their blades giving the necessary figures, they may almost be considered as standard. The tables may not be arranged in the same manner on all these squares, and on some they may be more complete than on others.

Under the heading *LENGTH COMMON RAFTERS PER FOOT RUN* (see Figure 17-31) will be found numbers, usually from 3 to 20. With each number is a figure in inches and decimal hundredths. The integers represent the rises of the rafters per foot of run, and the inches and decimals represent the lengths of the rafters per foot of run. As an example of the use of these tables, take a building 28 feet, 2 inches wide, thereby making the run of the rafters 14 feet (allowing for a 2-inch ridge). Let the desired pitch be 4 inches per foot. Under the number 4 : on the square will be found the length per foot of run—12.64 inches. The calculation for the length of the rafter is as follows:

Figure 17-31 The length-per-foot-run tables on one type of rafter framing square.

$$12.64 \times 14 = 176.96 \text{ inches}$$

$$\frac{176.96}{12} = 14.75 \text{ feet, or } 14 \text{ feet, } 9 \text{ inches}$$

If the run is in feet and inches, it is most convenient to reduce the inches to the decimal parts of a foot, according to Table 17-9:

Table 17-9   Reducing Inches to Decimal Parts of a Foot

Inches	*Part of Foot*
1	0.083
2	0.167
3	0.250
4	0.333
5	0.417
6	0.500
7	0.583
8	0.667
9	0.750
10	0.833
11	0.917

Problem 25
Find the length of a hip or valley rafter having an 8-inch rise per foot on a 20-foot building with the run of the common rafters measuring 10 feet.

Look for *LENGTH OF HIP OR VALLEY RAFTERS PER FOOT RUN* (see Figure 17-31), and read under the 8-inch rise the figure 18.76. This is the calculation:

$$18.76 \times 10 = 187.6 \text{ inches}$$

$$\frac{187.6}{12} = 15 \text{ feet, } 7.6 \text{ inches}$$

One edge of all good steel squares is divided into tenths of inches, so this length may be measured off directly on the rafter pattern with the steel square.

Problem 26
Find the difference in lengths of jack rafters on a roof with an 8-inch rise per foot and with a spacing of 16 inches on centers.

Under *DIFFERENCE IN LENGTH OF JACKS* (16-inch centers) on the square find the figure 19.23 below the figure 8 (rise per foot of run). This is the length of the first jack rafter, and the length of each succeeding jack will be 19.23 inches greater—38.46 inches, 57.69 inches, 76.92 inches, and so on.

Problem 27
Find the side cuts of jacks on a square (see Figure 17-32).

The fifth line is marked *SIDE CUT OF JACKS USE THE MARKS* ∧. If the rise is 8 inches per foot, find the figure 10 under figure 8 in the upper line. The proper side cut will then be 10 × 12, cut on 12. The side cuts for hip or valley rafters are found in the sixth line. For the 8 × 12 roof, it is $10^7/_8$ × 12, cut on 12.

No discussion of rafter framing is complete without an explanation of one of the oldest and most useful (though probably not the most accurate) methods of laying out a rafter with a steel square. Any square may be used if it has legible inch marks representing the desired pitch. It is the same method used for the layout of stairs. The layout for a rafter with a 9-foot run has a pitch of 7 inches × 12 inches, making the rise of the rafter 5 feet 3 inches (see Figure 17-33). The steel square is applied nine times. Carefully mark each application, preferably with a knife. A hip rafter is laid out in exactly the same manner by using 17 instead of 12 in the run and applying the square nine times, as was done for the common rafter. For short rafters, this is probably the least time-consuming of any method.

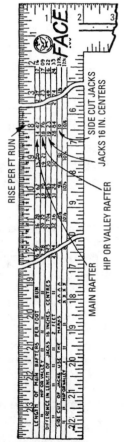

Figure 17-32 Typical rafter tables.

Table of Octagon Rafters
The complete framing square is provided with a table for cutting octagon rafters, as shown in Figure 17-34. In this table, the first line

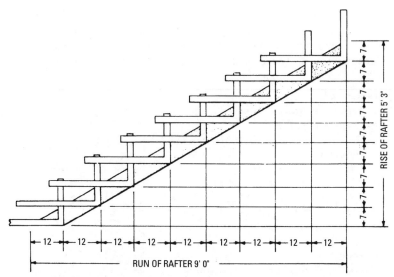

Figure 17-33 The method of stepping off a rafter with a square. The square is applied in consecutive steps (hence the name of the method).

of figures from the top gives the length of octagon hip rafters per foot of run. The second line of figures gives the length of jack rafters spaced 1 foot from the octagon hip. The third line of figures refers to the graduated edge that will give the side cut for octagon hip rafters. The fourth line of figures refers to the graduated edge that will give the side cuts for jack rafters. The tables are used in a manner similar to that used for the regular rafter tables just described and, therefore, need no further explanation. The last line (or bottom row of figures) gives the bevel of intersecting lines of various regular polygons. At the right end of the body on the bottom line can be read *MITER CUTS FOR POLYGONS—USE END OF BODY.*

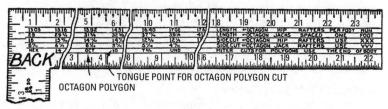

Figure 17-34 Typical octagon rafter tables.

Problem 28
Find the angle cut for an octagon.
For a figure of 8 sides, look to the right of the word *OCT* in the last line of figures, and find 10. This is the tongue reading. The end of the body is the other point, as shown in Figure 17-35.

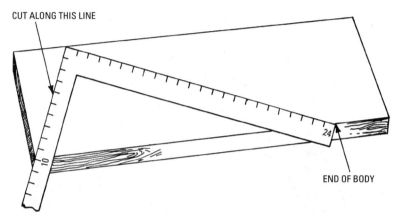

CUT ALONG THIS LINE

END OF BODY

Figure 17-35 The square in position for marking an octagon cut; it is set to point 10 on the tongue and to point 24 on the body.

Table of Angle Cuts for Polygons

This table is usually found on the face of the tongue. It gives the setting points at which the square should be placed to mark cuts for common polygons that have from 5 to 12 sides.

Problem 29
Find the bevel cuts for an octagon.
On the face of the tongue (see Figure 17-36), look along the line marked *ANGLE CUTS FOR POLYGONS,* and find the reading

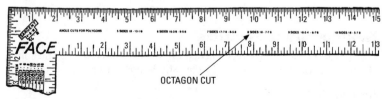

OCTAGON CUT

Figure 17-36 Table of angle cuts for polygons on the face of the square.

"8 sides 18–7½." This means that the square must be placed at 18 on one arm and at 7½ on the other to obtain the octagon cut (see Figure 17-37).

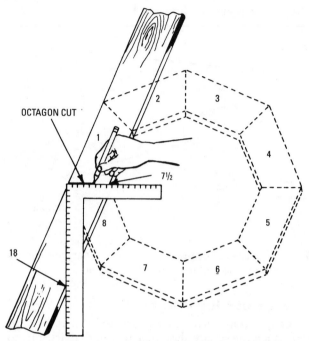

Figure 17-37 The application of the square for making angle cuts of polygons. The square is shown set to points 18 and 7½. When constructing an 8-sided figure, such as an octagon cap, the last figure in the reading is the setting for making the side. Mark as shown. Cut eight pieces to equal length, with this angle cut at each end of each piece. The pieces will fit together to make an 8-sided figure, as shown by the dotted lines.

Table of Brace Measure

This table on the square (see Figure 17-38) is located along the center of the back of the tongue and gives the length of common braces.

Problem 30

If the run is 36 inches on the post and 36 inches on the beam, what is the length of the brace?

In the brace table along the central portion of the back of the tongue (Figures 17-38 and 17-39), look at L for the following:

   36

    50.91

   36

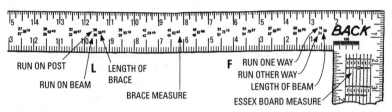

Figure 17-38   Table of brace measure on the back of the square.

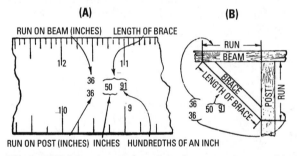

Figure 17-39   A portion of the brace-measure table, with an explanation of the various figures, is shown in A. The brace in position, illustrating the measurements of the brace-measure table, is shown in B.

This reading means that for a run of 36 inches on the post and 36 inches on the beam, the length of the beam is 50.91 inches.

At the end of the table (at F near the body) will be found the following reading:

   18

    30

   24

This means that where the run is 18 inches one way and 24 inches the other, the length of the brace is 30 inches.

The best way to find the length of the brace for the runs not given on the square is to multiply the length of the run by 1.4142 feet (when the run is given in feet) or by 16.97 inches (when the run is given in inches). This rule applies only when both runs are the same.

Octagon Table or Eight-Square Scale

This table on the square is usually located along the middle of the tongue face and is used for laying off lines to cut an eight-square or octagon-shaped piece of timber from a square timber.

In Figure 17-40, let ABCD represent the end section (or butt) of a square piece of 6-inch × 6-inch timber. Through the center, draw the lines AB and CD parallel with the sides and at right angles to each other. With dividers, take as many squares (6) from the scale as there are inches in width of the piece of timber, and lay off this square on either side of the point A, such as Aa and Ah. Lay off in the same way the same spaces from the point B (such as Bd and Be). Also lay off Cb, Cc, Df, and Dg. Then draw the lines ab, cd, ef, and gh. Cut off at the edges to lines ab, cd, ef, and gh, thus obtaining the octagon or 8-sided piece.

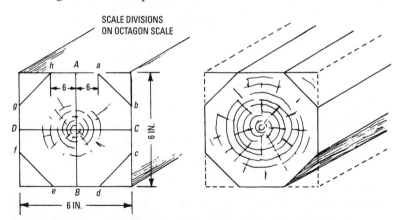

Figure 17-40 A square timber and its appearance after it has been cut to an octagonal shape, which shows the application of the octagon scale.

Essex Board Measure Table

This table is illustrated in Figure 17-41. It normally appears on the back of the tongue on the square. To use the table, the inch graduations on the outer edge of the square are used in combination with the values along the five parallel lines. Measure the length and width of the board, and look under the 12-inch mark for the width in inches. Then follow the line on which this width is stamped toward either end until the inch mark is reached on the edge of the square where the number corresponds to the length of the board in feet. The number found under that inch mark will be the length of the board in feet and tenths. The first number is feet, and the second is tenths of feet.

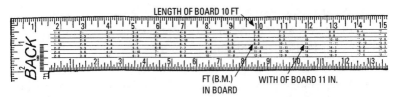

Figure 17-41   Table of Essex board measure on the back of the square.

Problem 31

How many feet Essex board measure in a board 11 inches wide, 10 feet long, and 1 inch thick? Also, 3 inches thick?

Find 11 under the 12-inch mark on the outer edge of the square (see Figure 17-41). That represents the width of the board in inches. Then follow on that line to the 10-inch mark (representing the length of the board in feet), and find on a line 9–2, which means that the board contains 9.2 feet board measure for a thickness of 1 inch. If the thickness were 3 inches, then the board would contain 9.2 feet × 3, or 27.6 B.M.

Summary

On most construction work (especially in house framing), the steel square is invaluable for accurate measuring and for determining angles.

The square most generally used has an 18-inch tongue and a 24-inch body. The body is generally 2 inches wide, and the tongue is $1\frac{1}{2}$ inches wide, varying in thickness from $\frac{3}{16}$ to $\frac{3}{32}$ of an inch. The various markings on the square are tables and scales or graduations.

Since the tables on the square relate mostly to problems encountered in cutting lumber for roof-frame work, it is necessary to know roof construction and the names of various rafters. These names are rise, run, rise per foot run, hip and valley rafters, jack and cripple rafters, common rafters, ridge, and plate.

The rafter tables vary considerably with different makes of squares, not only in the way they are calculated, but also in their positions on the square. Some tables are found on the face of the body, and others are on the back of the body. The two general classes of rafter tables found on squares are length of rafter per foot of run, and total length of rafter.

Review Questions

1. It is called a steel square, but what is the correct name for this tool?

2. What type of tables are found on the body of the square?

3. Name the various types of roof rafters.

4. What is rise per foot run of a roof?

5. What is rafter pitch?

6. When buying a square, it is advisable to get one with all the markings, rather than a _____ unit.

7. The term _____ is used to denote the inch divisions of the tongue and body length found on the outer and inner edges of the square.

8. The _____ of a common rafter is the distance found by following a plumb line from a point on the centerline of the top of the ridge to the level of the top of the plate.

9. The _____ per foot is the basis on which rafter tables and some squares are made.

10. To obtain the pitch of any roof, divide the rise of the rafter by the _____ the run.

Index

A

accidents, avoiding
  bench grinder, 194
  circular saw, 48–49
  clothing, proper, 46–47
  electrical drill, 50–51
  emergency surgery for saw
      accidents, 53–54
  first aid, 53–54
  during lifting and carrying, 53
  protective gear, 47, 171
  radial arm saw, 49–50
  table saw, 49
  for tools, hand and power,
      47–51, 181–182
  in worksite environment,
      51–53
adjustable wrench, 175, 176
adze, 128–129
  grinding, 193
aligned dimensioning, 24
aliphatic glue, 211
aluminum oxide, 153
alundum, 199
angle bridle joint, 254
angles, measuring and laying out
  with bevel and framing square,
      69
  of dovetailing, 248–249
  with framing square, 281–284
  of mitered moldings, 268–271
  of polygons, 311–312
  pitch of roof, 290, 291
  with protractor, 13, 15
  table of, 283
  with triangles, 16–20
arcs, drawing, 19, 21–22
Arkansas oilstone, 199
art gum eraser, 8
artificial oilstones, 199–200
augers, 157–161
  bits of, 160
  expansive bit, 161
  heads, types of, 160
  sharpening, 159

axe, 125, 128
  grinding, 193

B

back-and-forth saw, 183, 185
back of framing square, 277
backsaw, 108
ball-peen hammer, 169
band clamp, 94, 96
bar clamp, 94
barefaced tenon joint, 233
barrel hatchet, 126
batteries, 179
battery-powered tools, 179–181
  advantages, 180
  disadvantages, 180–181
  portable drills, 180, 181, 182,
      183, 184
belt sander, 186
benches, 98–105
  features of, 100–105
  portable, 100
  shop bench, 99
  temporary site bench, 99, 100
  types of, 98–100
  vises located upon, 97–98, 101,
      103
bench grinder, 191, 193–194
  safety, 194
bench hook, 104–105
bench plane, 136
bench rule, 82
bench stop, 103–104
bench vise, 97–98, 101, 103
bending forces, 229–231
bending scarf joint, 229–231
beveled joints, 211–212
  dovetail, 250
bevel of plane iron, proper, 140,
      141
bevel, sliding T, 67, 69
biscuits for plate jointery, 260
bit, 157–161
  machine-spur, 159
  router, 188

bit (*continued*)
 solid-center, 159
 types of, 160, 161
bit gage, 159, 161
blade (body) of framing square,
 62, 277
blind (secret) dovetail joint, 246
 halved, 253
block plane, 133, 135–136
 low-angle, 136
 use of, proper, 147, 148
board rule, 84–86
body (blade) of framing square,
 62, 277
boring tools, 157
 augers, 157–161
 awls, 157–158
 countersinks, 162
 twist bits, 162
bow compass, 7, 22
bow divider, 7
box-end wrench, 174–175
brace, for boring tools, 157
 hand, 163–165
 ratchet, 165
brace table, 64
brad awl, 157
 proper use of, 158
breathing protection, 47
bridle (open-tenon) joints, 253,
 254
broad hatchet, 125, 126
bull-nose plane, 143, 144
burnishers, proper use of,
 150–151, 152
butt chisel, 117–128, 120
butt joint, 213–219
 doweled, 215–218
 straight, 214–215
butt marking gage, 88–89

C

cabinetmaking, 209
 barefaced tenon joint, 233
 beveled joint, 211–212
 bridle (open-tenon) joint, 253,
 254

butt joint, 213–219
carcass construction, 255–256
characteristics of mortice-and
 tenon joints, 231
coopered joint, 213, 214
coped joints, 274–275
corner joint, 218–219, 245
cutting mortise-and-tenon joints,
 238–242
dado joint, 227, 228
degree of mortice housing, 234,
 235
double tenon joint, 234
dovetail angles, determining,
 248–249
dovetail joint, 243–252
dowel joint, 215–218
draw boring, 242–243
fastening the tenon, methods of,
 236
framed construction, 254–255
framing joints, 256, 257
glued joint, 211
halved joint, 252–254
hidden slot screwed joint,
 212–213
joints and joint designs,
 209–210
lap joint, 226
layout of mortice-and-tenon
 joints, 237–238
long and short shoulder tenon
 joint, 233, 234
mitered joint, 219–222, 223
mortice-and-tenon joint, types
 of, 231–234
plain edge joint, 213–219
plate jointery, 256–261
position of tenon, 234, 235
rabbeted (haunched) tenon joint,
 232, 233
rabbet joint, 226–227
scarf joint, 226, 227–231
splice joint, 224–226
splined joint, 222–224
straight joint, 213–215,
 218–219, 244

stub tenon joint, 231–233
tongue-and-groove joint, 248
types of joints, 210
caliper rule, 82–83
cap of double plane iron, 143
   sharpening of, proper, 144, 146
carborundum oilstones, 121–122,
   199
   double, 195
carcass construction, 255–256
carpenter's glue, 211
carpenter's light blade, 125
carpenter's pencil, 75–76
carpenter's razor blade, 125
carriage makers' narrow blade,
   125
carrying heavy objects, 53
C-clamp, 91
center bit, 161
centerlines, 25
center square, 70
central processing unit, 26
chalk box, 75, 76, 77
charger, battery, 179
chisels, 115–124
   butt chisel, 117–118, 120
   cutting mortices with, 239–240
   firmer chisel, 116, 117
   framing chisel, 116, 117
   gouges, 117, 119
   honing, 196–198
   mill chisel, 117–118, 120
   mortise chisel, 116, 117
   paring chisel, 116
   pocket chisel, 117–128, 120
   selecting and using, 118–121,
      122, 123
   sharpening, 121–124
   slick, 117
   socket chisel, 117, 119
   tang chisel, 117, 119
circles, drawing, 19, 21–22
clamp-on vises, 98
clamps
   band clamp, 94, 96
   bar clamp, 94
   C-clamp, 91

deep-throat clamp, 92
   edge clamp, 94
   hand screw clamp, 94, 95
   pipe clamp, 94
   spring clamp, 94, 95
   steel-bar clamp, 94
claw hatchet, 126
cleaning powder, 13, 14
clothing, 46
coach makers' razor blade, 125
coated abrasives, 153
combination square, 66–67, 68
common (plain, through) dovetail
   joint, 244, 245, 250–251
common (main) rafter, 289
compass, 6–7, 79
   extension bar, 6, 7
   sharpening lead of, 12
   use of, proper 19, 21–22, 23
compass saw, 108
component graphics object, made
   from primitives, 37, 40–41
compound dovetail joint, 244–246
compression forces, 225–226,
   228–229, 231
compression scarf joint, 229–231
computer-aided drafting (CAD)
   benefits of using, 29–30, 37
   commands to drawing software,
      35, 36–41
   equipment required for, 26–27,
      30–31, 41–42
   example of drawing using, 35–41
   software required for, using,
      27–28, 31–41
computer system, personal, 26
concave-bottom spoke-shave,
   132
contact cement, 211
convex-bottom spoke-shave, 132
coopered joints, 213, 214
coping, 274–275
coping saw, 109, 111, 113
cordless drill, 180, 181
corner joint, 218–219
   dovetailed, 245
cosines, table of, 283

countersink bit, 161, 162
countersinking tools. *See* boring
    tools
cripple rafter, 295
crosscut saw, 107, 109–111
    sharpening, 202–203, 205–206
    teeth and action of, 109–111
    use of, proper, 111
cross-lap joint, 226
cross-peen hammer, 169
cursor, 31, 35
curved-claw hammer, 167
cutting tools
    chisels, 115–124
    drawknives, 124–125
    hand saws, 107–113
    hatchets and axes, 125–129
    See also smoothing tools

D

dado joint, 227, 228
databases of drawings, 29–30, 33
data files, 27, 31, 33–34
data processing, 28
deck of roof, 290
deep-throat clamp, 92
depth gage, 159, 161
diagonal scale, 62
digitizer tablet, 31, 35
dimension drawings, 24–25
dimension lines, 24–25
diminished dovetail joint, 252
dividers, 7
    spacing with, 22–24
    winged, 79
dot-matrix printer, 31
double-bar marking gage, 86–87
double plane iron, 141
    cap of, 143, 144, 146
double straightedge, 55
    testing surface winding with, 58
double-spur bit, 161
double tenon joint, 234
dovetail halved joint, 253
dovetail saw, 109
dovetail joints, 243–252
    angles of, finding, 248–249

beveled, 250
common (plain, through)
    dovetail, 244, 245,
    250–251
compound dovetail, 244–246
diminished dovetail, 252
halved, 253
lap (half-blind) dovetail, 246,
    250–251
layout template for, 249
mortice (blind) dovetail, 246
pin of, 244
socket of, 244
dovetail template, 249
dowel joint, 215–218
    method of making, 216–217
    See also plate jointery
drafting, 1–2
    arcs and circles, 19, 21–22, 23
    compared with computer-aided
        drafting, 29
    computer-aided (*see*
        computer-aided drafting)
    dimensioning, 24–25
    instruments, 1–13, 14, 27
    spacing, 22–24
    straight lines, 15–19, 21
drafting, computer-aided (CAD)
    benefits of using, 29–30, 37
    commands to drawing software,
        35, 36–41
    equipment required for, 26–27,
        30–31, 41–42
    example of drawing using,
        35–41
    software required for, using,
        27–28, 31–41
drafting machine, 5
drafting table, 3, 4
draw boring, 242–243
drawing boards, 3
drawing databases, 29–30, 33
drawing instruments, 1–13, 14,
    27
    basic set of, 6
drawing nails out with hammer,
    171

drawing pencils, 12–13
  hardness of, 12
  sharpening, 12–13
drawing primitives, grouping into
    components, 37, 40–41
drawings, databases of, 29–30, 33
drawing, technical, 1–2
  arcs and circles, 19, 21–22, 23
  compared with computer-aided
      drafting, 29
  computer-aided (see
      computer-aided drafting)
  dimensioning, 24–25
  instruments, 1–13, 14, 27
  spacing, 22–24
  straight lines, 15–19, 21
drawknives, 124–125
  grinding, 193
dressing saws, 208
drill, electrical, 180, 181, 182,
    183, 184
  safety, 50–51
drill, hand, 163–165
  proper use of, 164–165
drilling tools. See boring tools
driving in nails with hammer,
    169–171
drum plotter, 42
dusting brush, 8
drywall saw, 109

E
edge clamp, 94
electronic drawing. See
    computer-aided drafting
    (CAD)
emery, 153
end lap joint, 226
enlarging tools. See boring tools
environment, safety of, 51–53
epoxy glue, 211
erasers, 8
Essex board measure table, 63–64,
    314–315
excavations, 51
expansive auger bit, 161
extension bar, compass, 6, 7

extension lines, 24–25
eye protection, 47, 171

F
face of framing square, 277
falling objects, 53
fastening tools, 167
  hammers, 167–171
  screwdrivers, 172–174
  staplers and guns, 176–177
  wrenches, 174–176
feathers, 222
files, 151–152
  types of, 153
filing saws, 205–208
firmer chisel, 116, 117
first-aid training, 53
fish joints. See splice joints
fish plates, 225
flat-bottom spoke-shave, 132
flint, 153
flush molding, 267–268
flute of bit, 159
folding wooden rule, 81
footwear, 46
fore plane, 133, 134
  shape of cutting edge, 142
Forstner bits, 160
framed construction, 254–255
frame miter joint, 220, 221
framing chisel, 116, 117
framing joints
framing square (steel square),
    62–66
  brace measures, 312–314
  laying off angles with, 281–284
  polygons, determining cuts for,
      311–312
  portions of, 277–278
  scale problems, solving with,
      278–284
  square-and-bevel problems,
      solving with, 284–289
  table (roof-frame) problems,
      solving with, 289–311
  timber and board measure,
      314–315

framing table, 62–63
french curves, 13, 14

G
gages, marking, 86–89
  butt gage, 88–89
  double bar gage, 86–87
  single bar gage, 86
  slide gage, 87–88
garnet, 153
glued joints, 211
glue gun, 211
glues, types of, 211
gouge chisels, 117, 119
  honing, 196–197
grades (grit size)
  of carborundum oilstones, 199
  of India oilstones, 200
  of sandpaper, 154
graphic input device, 30–31,
    34–35
grinder, bench, 191, 193–194
  safety, 194
grinding, 191–194
  plane irons, 144–145
grooves in plate jointery, cutting,
    258–259
grooving plane, 137
guiding tools
  depth gage, 159, 161
  miter box, 265–266
  straightedge, 55–56

H
hacksaw, 109
hair-spring divider, 7
  spacing with, 22–24
half-blind (lapped) dovetail joint,
    246, 250, 251
half hatchet, 126
half-lap joint, 226. *See also* scarf
    joint
halved joints, 252–254
halved lap joint, 253
halved tee joint, 253
hammer, 167–171
  driving nails with, 169–171

types of, 168–170
  use of, proper, 169–171
hand axe, 125, 127
hand brace, 163–165
hand drill, 163–165
handle scraper, 148
hand reconstructive surgery,
    53–54
hand saw
  backsaw, 108
  compass saw, 108
  coping saw, 109, 111, 113
  crosscut, 107, 109–111
  dovetail saw, 109
  dressing of, 208
  drywall saw, 109
  hacksaw, 109
  keyhole saw, 108
  panel, 108
  ripsaw, 107, 110–111, 112,
      113
  rod saw, 109
  sharpening, 201–208
  sizes of, 108
  teeth and action of, 109–111
  types of, 107–109
hand screw, 94, 95
hand tool safety, 47–48
hard Arkansas oilstone, 199
hard hat, 47
hardware, 26
hatchets, 125–129
  adze, 128–129
  axe, 125, 128
  hand axe, 125, 127
  sharpening, 127, 129, 193
  types of, 126
  use of, proper, 127, 129
haunched (rabbeted) tenon joint,
    232, 233
head protection, 47
heavy plane iron, 139
heel cut, 291, 292
heel of framing square, 62
hidden slot screwed joints,
    212–213
hip rafter, 291, 294

holding tools, 91
  clamps, 91–96
  horses, 91, 92, 93
  vises, 96–98
  workbenches, 98–105
holes, boring, 164–165
hollow-bottom spoke-shave, 132
horses, 91, 92, 93
  construction of, 92, 93
hot-melt glue, 211
housing of mortice
  degree of, 234, 235
  dovetailed, 250–252
hundredths scale of framing
    square, 64

I
inch scale of framing square, 64
India oilstones, 199–200
inside caliper, 83
inside miter, 273

J
jack plane, 133, 134, 142
  shape of cutting edge, 139–141,
    142
jack rafter, 295
jointer plane, 134, 142
  shape of cutting edge, 140–141
jointing saw teeth, 202
joint, dovetail, 243–252
  beveled, 250
  common (plain, through)
      dovetail, 244, 245, 250–251
  compound dovetail, 244–246
  diminished dovetail, 252
  finding angles of, 248–249
  halved, 253
  lap (half-blind) dovetail, 246,
    250–251
  layout template for, 249
  mortice (blind) dovetail, 246
  pin of, 244
  socket of, 244
joint, mortice-and-tenon
  barefaced tenon joint, 233
  characteristics of, 231

cutting the mortice, 238–240
cutting the tenon, 240–242
degree of mortice housing, 234,
    235
double tenon joint, 234
draw boring, 242–243
fastening the tenon, methods of,
    236
laying out, 237–238
long and short shoulder tenon
    joint, 233, 234
position of tenon, 234, 235
rabbeted (haunched) tenon joint,
    232, 233
stub tenon joint, 231–233
types of, 231–234
joints and joint designs, 209–210
  beveled joint, 211–212
  bridle (open-tenon) joint, 253,
    254
  butt joint, 213–219
  coopered joint, 213, 214
  coped joints, 274–275
  corner joint, 218–219, 245
  cutting mortise-and-tenon joints,
    238–242
  dado joint, 227, 228
  dovetail angles, determining,
    248–249
  dovetail joint, 243–252
  dowel joint, 215–218
  framing joints, 256, 257
  glued joint, 211
  halved joint, 252–254
  hidden slot screwed joint,
    212–213
  lap joint, 226
  layout of mortice-and-tenon
    joints, 237–238
  mitered joint, 219–222, 223
  mortice-and-tenon joint (*see*
      mortice-and-tenon joints)
  plain edge joint, 213–219
  plate jointery, 256–261
  rabbet joint, 226–227
  scarf joint, 226, 227–231
  splice joint, 224–226

joints and joint designs
(*continued*)
splined joint, 222–224
straight joint, 213–215,
218–219, 244
tongue-and-groove joint, 248
types of joints, 210

K
kerf, 110–111
keyboard, 26
keyhole saw, 108

L
ladder safety, 52
lap joints, 226
dovetailed (half-blind), 246, 250,
251
lath hatchet, 126
laying off angles
with bevel and framing square,
69
of dovetailing, 248–249
with framing square, 281–284
of mitered moldings, 268–271
of polygons, 311–312
pitch of roof, 290, 291
with protractor, 13, 15
table of, 283
with triangles, 16–20
laying out, 75
layout tools
awl, 76
chalk box, 75
compass, 79
dividers, 79
marking gages, 86–89
pencils, 75–76
leader lines, 25
lengthening bar, compass, 6, 7
level, spirit, 70, 71
lifting heavy objects, 53
Lily White Washita oilstone,
198
lines (drawn), 15–19
dimensioning, styles of, 24–25
oblique, 16–20

long and short shoulder tenon
joint, 233, 234
low-angle block plane, 136
lumber scale, 84–86

M
machine-spur bit, 159
machinist's vise, 97–98
main (common) rafter, 289
marking awl, 76, 78
marking gages, 86–89
butt gage, 88–89
double bar gage, 86–87
single bar gage, 86
slide gage, 87–88
marking tools. *See* layout tools
measuring angles
with bevel and framing square,
69
of dovetailing, 248–249
with framing square, 281–284
of mitered moldings, 268–271
of polygons, 311–312
pitch of roof, 290, 291
with protractor, 13, 15
table of, 283
with triangles, 16–20
measuring tape, steel, 81
measuring tools
rules, 81–86
mechanical drawing, 1–2
arcs and circles, 19, 21–22, 23
compared with computer-aided
drafting, 29
computer-aided (*see*
computer-aided drafting)
dimensioning, 24–25
instruments, 1–13, 14, 27
spacing, 22–24
straight lines, 15–19, 21
mechanical pencils, 12, 13
memory, of computer system, 27
middle lap joint, 226
mill chisel, 117–118, 120
mill file, 151
miter, 59
miter angle, 268

miter box, 221, 265–266
mitered bridle joint, 254
mitered halved joint, 253
mitered joint, 219–222, 223
  constructing, 221–222
mitering, 265–275
  cutting long miters, 273–274
  tools for, 265–267
miter saw, 265–267
miter square, 59
molding plane, 133
moldings
  coping, 274–275
  cutting, 267–273
monkey wrench, 175–176
mortice, 231, 232
  cutting, 238–240
  housing, degree of, 234, 235
mortice-and-tenon joints
  barefaced tenon joint, 233
  characteristics of, 231
  cutting the mortice, 238–240
  cutting the tenon, 240–242
  degree of mortice housing, 234,
    235
  double tenon joint, 234
  draw boring, 242–243
  fastening the tenon, methods of,
    236
  laying out, 237–238
  long and short shoulder tenon
    joint, 233, 234
  position of tenon, 234, 235
  rabbeted (haunched) tenon joint,
    232, 233
  stub tenon joint, 231–233
  types of, 231–234
mouse, 31, 35
mouth of plane, 143, 144
movable-head T-square, 7–8
multispur bit, 161

N
nailing gun, 177
nails
  driving in, with hammer,
    169–171

  drawing out, with hammer,
    171
National Electrical Code, 49
natural oilstones, 198–199
No. 1 Washita oilstone, 198

O
oblique halved joint, 253
oblique bridle joint, 254
oblique lines, drawing, 16–20
Occupational Safety and Health
    Administration (OSHA), 47
octagon rafter, 296–297,
    309–310
octagon scale, 62
offset screwdriver, 174
oilstones, 194–200
  cleaning, 196
  double carborundum, 195
  types of, 198–200
  use of, proper, 145, 195–198
open-end wrench, 174, 175
open-tenon (bridle) joints, 253,
    254
orbital sander, 186
out of true, 60
outside caliper, 83
outside miter, 273

P
panel saw, 108
parallel straightedge, 3, 4
paring chisel, 116
pencils, 12–13, 76
  carpenter's, 75–76
  hardness of, 12
  mechanical, 12, 13
  sharpening, 12–13, 78
  use of, correct, 16, 21, 24, 56,
    78
personal computer system, 26
Phillips head screwdriver,
    173–174
picking and placing graphics
    objects, 35, 36–41
picture-frame vise, 220
pilasters, fastening, 213

pin of dovetail joint, 244
  spacing and position of,
    246–248
pipe clamp, 94
pitch of roof, 290, 291
plain (common, through) dovetail
  joint, 244, 245, 250–251
plain dividers, 7
plain edge joint, 213–219
  doweled, 215–218
  straight, 214–215
plane iron (plane cutter), 132
  bevel of cutting edge, 140, 141
  double, 141, 143
  grinding, 144–145, 192–193
  honing, 196–198
  set of, 146–147
  shape of cutting edge, 139–141,
    142, 192–193
  single, 140, 143
  types of, 138
  whetting, 144, 145
planes, 132
  bench plane, 136
  block plane, 133, 135–136
  bull-nose plane, 143, 144
  electric, 183–184, 186
  fore plane, 133, 134
  irons for, 138–143
  jack plane, 133, 134
  jointer plane, 134
  molding plane, 133
  mouth of, 143, 144
  plow plane, 137
  rabbet plane, 136
  router plane, 137–138
  sharpening of, 144–147
  smoothing plane, 133,
    134–135
  Surform plane, 136–137
  trenching plane, 137
  use of, proper, 143–147, 148
planing technique, 147, 148
plate
  fish plate, 225
  in joint, 260
  of roof, 289

plate jointery, 256–261
  advantages and disadvantages of,
    260
plotter, 31, 41–42
plow plane, 137
plumb, 70
plumb bob, 70–72
plumb cut, 291, 292
pocket chisel, 117–118, 120
pocket scriber, 76, 78
pointing device, 30–31, 34–35
polyvinyl glue, 211
portable workbenches, 100
power tools, electrical
  cordless, battery-powered
    179–181
  drills, 180, 181, 182, 183, 184
  groove cutters for plate jointery,
    258–259
  planes, 183–184, 186
  routers, 186, 188–189
  sanders, 186, 187–188
  safety, 48–51, 181–182
  saws, 182–183, 184–186, 187
printer, 27, 31
protective gear, 47, 171
protractors, 13, 15
  steel precision, 15
punching tools. *See* boring tools

R
rabbet and round joint, 257
rabbeted tenon joint, 232, 233
rabbet plane, 136
rabbet joint, 226–227
rafters, 289
  common (main) rafter, 289
  cripple rafter, 295
  heel cut of, 291, 292
  hip rafter, 291, 294
  jack rafter, 295
  octagon rafter, 296–297,
    309–310
  plumb cut of, 291, 292
  valley rafter, 293, 294
rafter tables, 62–63, 300–310
rake molding, 268–271

rasps, 151–152, 153
  types of, 153
ratchet brace, 165
ratchet screwdriver, 172–173, 174
reciprocating sander, 186
reciprocating saw, 183, 185
Red Cross, 53
respirator, 47
retaining tools, 91
  clamps, 91–96
  vises, 96–98
return bead and butt joint, 257
reuse of electronic drawing
    databases, 29–30
ridge of roof, 289
rise per foot run, 290
ripping hammer, 167
ripsaw, 107, 110–111, 112, 113
  sharpening, 202–203, 207–208
  teeth and action of, 110, 111,
    112
  use of, proper, 111, 112, 113
rivet gun, 176
riveting hammer, 170
rod saw, 109
roof-frame construction
  common (main) rafter, 289
  deck, 290
  hip rafter, 291
  pitch, 290, 291
  plate, 289
  ridge, 289
  rise per foot run, 290
  span, 290
roof-frame problems, 289–311
rose countersink, 162
Rosy Red Washita oilstone,
    198
router, electrical, 186, 188–189
router plane, 137–138
ruby eraser, 8
rules, 10, 11, 81–86
  bench rule, 82
  board rule, 84–86
  caliper rule, 82–83
  folding wooden rule, 81
  use of, proper, 83–84

S
saber saw, 184–186, 187
saddle tree shave, 125
safety, 45–46
  bench grinder, 194
  circular saw, 48–49
  clothing, proper, 46–47
  electrical drill, 50–51
  emergency surgery for saw
    accidents, 53–54
  first aid, 53–54
  during lifting and carrying, 53
  protective gear, 47, 171
  radial arm saw, 49–50
  table saw, 49
  for tools, hand and power,
    47–51, 181–182
  in worksite environment, 51–53
safety goggles, 47
sanders, portable, 186, 187–188
sandpaper, 152–154
  types of, 154
saw accidents, emergency
    treatment for, 53–54
saw blade sharpening, 201–208
  dressing, 208
  filing, 205–208
  jointing, 202
  setting, 204
  shaping, 202–204
saw, circular, 182–183, 185
  safety with, 48–49
saw, hand
  backsaw, 108
  compass saw, 108
  coping saw, 109, 111, 113
  crosscut, 107, 109–111
  dovetail saw, 109
  dressing of, 208
  drywall saw, 109
  hacksaw, 109
  keyhole saw, 108
  panel, 108
  ripsaw, 107, 110–111, 112, 113
  rod saw, 109
  sharpening, 201–208
  sizes of, 108

saw, hand (*continued*)
  teeth and action of, 109–111
  types of, 107–109
sawhorses, 91, 92, 93
  construction of, 92, 93
sawhorse vise, 98
saw, miter, 265–267
saw, power, 182–183, 184–186, 187
  reciprocating saw, 183, 185
  saber saw, 184–186, 187
saw, radial arm
  safety, 49–50
saw set, 204
saw, table
  safety with, 49
saw teeth, 109–112, 113
  adjusting set of, 204
  angles of, 111, 113
  jointing, 202
  set of, 109, 204
  shaping, 202–204
  sharpening, 205–208
scaffolding, safety of, 51–52
scale drawing, 24–25, 32
scale problems, 278–284
scales, 10, 11
  lumber scale, 84–86
scarf joint, 226, 227–231
scraper-plane, 148
scrapers, 147–151, 152
  sharpening, 150–151, 152
  types of, 148
  use of, proper 150, 151
scraper steel, proper use of, 150–151, 152
scratch awl, 76, 78, 157, 158
screen, 27
screwdrivers, 172–174
  Phillips, 173–174
  ratchet, 172–173, 174
screwed miter joint, 222, 223
scriber, 76, 78
secret (blind) dovetail joint, 246
set of plane iron, 146–147
set of saw teeth, 109
  adjusting, 204

shank, 157
shaping saw teeth, 202–204
sharpening of tools, 191–200
  adzes, 193
  augers, 159
  axes, 127, 129, 193
  chisels, 121–124
  drawknives, 193
  grinding step, 191–194
  hatchets, 193
  honing guide, use of, 146
  honing step, 194–198
  pencils, 12–13, 78
  planes, 144–147, 192–193
  saws (*see* saw blade sharpening)
  scrapers, 150–151, 152
  spoke-shaves, 193
shingle shave, heavy blade
shingling hatchet, 126
shoes, 46
shop bench, 99
silicon carbide, 153, 154
sines, table of, 283
single-bar marking gage, 86
single-pin (plain) dovetail joint, 244, 245
single plane iron, 140, 143
single straightedge, 55
slick, 117
slide marking gage, 87–88
slip-tongue joint. *See* splined joint
smoothing plane, 133, 134–135
  shape of cutting edge, 140–141
smoothing tools, 131
  files and rasps, 151–152
  planes, 132–147, 148
  sandpaper, 152–154
  scrapers, 147–151
  spoke-shaves, 131–132
socket chisel, 117, 119
socket of dovetail joint, 244
  spacing and position of, 246–248
soft Arkansas oilstone, 199
soft-faced hammer, 170
software, 27–28
solid-center bit, 159

span of roof, 290
spar shave, 125
spirit level, 70, 71
splayed corner joint, 257
splice joint, 224–226
splined joint, 222–224
  construction of, 224
splines, constructing, 224, 225
spoke-shave, 131–132
  grinding, 193
  proper use of, 131
spring clamp, 94, 95
spring molding, 268–271
square-and-bevel problems,
    284–289
squares
  center square, 70
  combination square, 66–67, 68
  framing square, 62–66 (*see also*
      framing square)
  miter square, 59
  try-and-miter square, 59–60
  try square, 57–59
staple gun, 176–177
steel-bar clamp, 94
steel measuring tape, 81
steel precision protractor, 15
steel square (framing square),
    62–66
  brace measures, 312–314
  laying off angles with, 281–284
  polygons, determining cuts for,
      311–312
  portions of, 277–278
  scale problems, solving with,
      278–284
  square-and-bevel problems,
      solving with, 284–289
  table (roof-frame) problems,
      solving with, 289–311
  timber and board measure,
      314–315
straight-claw hammer, 167
straightedge, 55–56
  drawing lines with, 56
  parallel, 3, 4
  testing surfaces with, 57, 58

straight joints, 213–215, 218–219
  dovetailed, 244
straight-peen hammer, 169
stub tenon joint, 231–233
supporting tools, 91
  horses, 91, 92, 93
  workbenches, 98–105
support pegs, 102–103
Surform plane, 136–137
syntax of commands, 35

T
table problems, 289–311
tang chisel, 117, 119
taper file, 151
tape, steel measuring, 81
technical drawing, 1–2
  arcs and circles, 19, 21–22, 23
  compared with computer-aided
      drafting, 29
  computer-aided (*see*
      computer-aided drafting)
  dimensioning, 24–25
  instruments, 1–13, 14, 27
  spacing, 22–24
  straight lines, 15–19, 21
tee bridle joint, 254
telescoping scriber, 76, 78
temporary site bench, 99
  construction of, 100
tenon, 231
  cutting, 240–242
tenon joint. *See* mortice-and-tenon
    joints
tension forces, 226, 228–229,
    230
tension scarf joint, 229
testing tools
  bevel, 67, 69
  combination square, 66–67, 68
  framing square, 62–66
  plumb bob, 70–72
  spirit level, 70
  straightedge, 55–56, 57, 58
  try square, 56–62
thin plane iron, 139
throat of bit, 159

through (common, plain) dovetail
  joint, 244, 245, 250–251
tongue-and-groove joint, 248
tongue of framing square, 62, 277
tonguing board, 224, 225
tool belt, xxi
tool drop, 103
tools, power. *See* power tools,
  electrical
tools, selecting and using, xix–xx
  chisels, 118–121, 122, 123
  framing squares, 278
  hammers, 168
  mechanical drawing instruments,
    2
tools, sharpening of, 191–200
  adzes, 193
  augers, 159
  axes, 127, 129, 193
  chisels, 121–124
  drawknives, 193
  grinding step, 191–194
  hatchets, 193
  honing guide, use of, 146
  honing step, 194–198
  pencils, 12–13, 78
  planes, 144–147, 192–193
  saws (*see* saw blade sharpening)
  scrapers, 150–151, 152
  spoke-shaves, 193
trenches, 51
trenching plane, 137
trestles, 91, 92, 93
  construction of, 92, 93
triangles, 10, 11
  drawing lines with, 15–19, 21
triangular scales, 10, 11
try-and-miter square, 59–60
try square (trying square),
    57–61
  testing squareness using,
    60, 61
  testing truth of, 60–61
T-square, 7–9
  drawing lines with, 15–19, 21
turning tools. *See* boring tools
twist bits, 160, 162

U
unidirectional dimensioning,
  24
unmounted scraper, 148, 149

V
valley rafter, 293, 294
varying miters, 268
video display, 27
vises, 96–98
  bench vise, 97–98
  clamp-on, 98
  machinist's vise, 97–98
  picture-frame vise, 220
  sawhorse vise, 98
  woodworker's vise, 96–97
  workbenches, located upon,
    97–98, 101, 103
visors, 47

W
wagon makers' heavy blade,
  125
Washita oilstone, 198
whetting of plane iron, 144,
  145
white glue, 211
winding, of surface, 58
winged dividers, 79
wire edge, 144, 145, 152
wood rasp, 152
woodworker's vise, 96–97
workbenches, 98–105
  features of, 100–105
  portable, 100
  shop bench, 99
  temporary site bench, 99, 100
  types of, 98–100
  vises, located upon, 97–98, 101,
    103
worksite safety, 51–53
  bench grinder, 194
  circular saw, 48–49
  clothing, proper, 46–47
  electrical drill, 50–51
  emergency surgery for saw
    accidents, 53–54

first aid, 53–54
during lifting and carrying, 53
protective gear, 47, 171
radial arm saw, 49–50
table saw, 49
for tools, hand and power,
  47–51, 181–182
in working environment, 51–53

wrenches, 174–176
adjustable, 174, 175
box-end, 174–175
monkey, 175–176
plain, 174, 175

X
X-Y plotter, 42